# CONSUMER ELECTRONICS

**AUTHORS**

**Dr. Amit M. Joshi**
*Assistant Professor,*
Malaviya National Institute of Technology Jaipur (MNIT), Jaipur

**Dr. Maulin M. Joshi**
*Assistant Professor,*
*Banasthali Vidyapith, Newai*

**Dr. Urvashi Prakash Shukla**
*Assistant Professor,*
Banasthali Vidyapith, Newai

**REVIEWER**

**Dr. Savitesh Madhulika Sharma**
*Professor and Dean (R&D),*
Siddartha Institute of Science and Technology,
Puttur (A.P)

KHANNA BOOK PUBLISHING CO. (P) Ltd.
4C/4344, Ansari Road, Darya Ganj,
New Delhi, India

All India Council for Technical Education
Nelson Mandela Marg, Vasant Kunj,
New Delhi, India

Dear Readers,

To prevent the piracy, this book is secured with HIGH SECURITY HOLOGRAM on the front title cover. In case you don't find the hologram on the front cover title, please write us to at contact@khannabooks.com or whatsapp us at +91-99109 09320 and avail special gift voucher for yourself.

Specimen of Hologram on front Cover title:

Moreover, there is a SPECIAL DISCOUNT COUPON for you with EVERY HOLOGRAM.
How to avail this SPECIAL DISCOUNT:
Step 1: Scratch the hologram
Step 2: Under the scratch area, your "coupon code" is available
Step 3: Logon to www.khannabooks.com
Step 4: Use your "coupon code" in the shopping cart and get your copy at a special discount
Step 5: Enjoy your reading!

**ISBN:** 978-93-5538-350-1
**Book Code:** DIP466EN

**Consumer Electronics**

*by* Dr. Amit M. Joshi, Dr. Maulin M. Joshi and Dr. Urvashi Prakash Shukla

**First Edition:** 2026

*Originally Published by:*
**All India Council for Technical Education, Ministry of Education, GOI**

*Exclusively Published & Distributed by:*
**Khanna Book Publishing Co. (P) Ltd.**
4C/4344, Ansari Road, Daryaganj,
New Delhi (INDIA)

Visit us at: **www.khannabooks.com**
Write us at: **contact@khannabooks.com**

To view complete list of books, Please scan the QR Code:
KPH

*Printed in Bharat.*

## UNDERGRADUATE COURSE BOOKS

| S.No. | ISBN | Book Name | Author |
|---|---|---|---|
| 1 | 9789355382160 | Advanced Electrical Machines: Theory and Practicals | Bharat Singh Rajpurohit |
| 2 | 9789355382139 | Analog Electronic Circuits: Theory and Practical | Venu Sonti, Sachin Jain |
| 3 | 9789355383808 | Analog Electronic Devices: Theory and Practicals | N. B Balamurugan |
| 4 | 9789391505158 | Basic Electrical Engineering (with Lab Manual) | S.K. Sahdev |
| 5 | 9789355388087 | Basic of Electronics | Brijesh Iyer |
| 6 | 9789355387127 | Basics of Electrical Machines: Theory and Practicals | D. K. Palwalia, U. K. Kalla, R. K. Kumawat |
| 7 | 9789355389381 | Basics of Mechanics | Shubhankar Bhowmick, N. Venkata Swamy Naidu |
| 8 | 9789355386250 | Basics of Thermodynamics | Parmod Kumar, Atul Dhar |
| 9 | 9789355386984 | Biology for Engineers | Annamma Odaneth |
| 10 | 9789391505141 | Chemistry-I | Manisha Agarwal |
| 11 | 9789355384843 | Civil Engineering Societal and Global Impact | Shakuntala Acharya |
| 12 | 9789355382184 | Computer Aided Civil Engineering Drawing | D. Chandra Mouli, B. Swathi |
| 13 | 9789355382580 | Computer Aided Design & Analysis | Hari Kumar Voruganti |
| 14 | 9789355387592 | Computer Architecture | D. Vaithiyanathan |
| 15 | 9789355385093 | Computer Networks | Hari Prabhat Gupta |
| 16 | 9789355385307 | Construction Engineering & Management | Sparsh Johari |
| 17 | 9789355388728 | Control Systems | S.N. Deepa |
| 18 | 9789355385420 | Data Structure and Algorithms | Ranbir Singh Sanasam |
| 19 | 9789355386700 | Design and Analysis of Algorithms | Hari P. Gupta, Rahul Mishra |
| 20 | 9789355388872 | Digital Electronics and System | Abhishek Bhatt |
| 21 | 9789355386571 | Digital Signal Processing | Sanjeev Sharma |
| 22 | 9789355383396 | Disaster management | Pradeep K. Goyal, Anil K. Gupta |
| 23 | 9789355385222 | Discrete Mathematical Structures | Narendra S. Chaudhari, Sharmila S. P. |
| 24 | 9789355387868 | Effective Technical Communication | Pritee Sharma, Sanchita Verma |
| 25 | 9789355383389 | Electrical Circuit Analysis and Network Theory | Shouri Chatterjee |
| 26 | 9789355388957 | Electromagnetic Waves | Uttam M. Pal, Srijith K. |
| 27 | 9789355384652 | Electronic Communication | Vimal Bhatia |
| 28 | 9789355388711 | Embedded Systems | Santanu Chattopadhyay |
| 29 | 9789355385147 | Energy Science and Engineering | M. Rizwan, Majid Jamil |
| 30 | 9789355387929 | Engineering Economics, Estimation & Costing | Yogesh C. Joshi |
| 31 | 9789355384669 | Engineering Geology | R. N. Khare, Ketan Chourasia |
| 32 | 9789391505066 | Engineering Graphics & Design | Pradeep Jain |
| 33 | 9789391505097 | English (with Lab Manual) | Kulbhushan Kumar |
| 34 | 9789355385963 | Environmental Engineering | Chandra Sekhar Matli |
| 35 | 9789355388117 | Fluid Mechanics | Shreenivas N. Londhe, Pradnya R. Dixit |
| 36 | 9789355388377 | Fluid Mechanics & Fluid Machines | R.P. Saini |
| 37 | 9789355387103 | Geotechnical Engineering | Paramita Bhattacharya |

## UNDERGRADUATE COURSE BOOKS

| S.No. | ISBN | Book Name | Author |
|---|---|---|---|
| 38 | 9789355385130 | Hydraulic Engineering | Sayantan Ganguly |
| 39 | 9789355385550 | Hydrology & Water Resources Engineering | Mitthan Lal Kansal, P.K. Agarwal |
| 40 | 9789355389688 | Information Technology Workshop in MATLAB | Hari Prabhat Gupta |
| 41 | 9789355383235 | Instrumentation for Civil Engineering Applications | Ashim Kanti Dey, Manita Das |
| 42 | 9789355387042 | Introduction to Civil Engineering | Manabendra Saharia, Nagendra R. Velaga |
| 43 | 9789355384805 | Introduction to Database Systems | Vikram Singh |
| 44 | 9789355387509 | Machine Element & System Design | Sabah Khan |
| 45 | 9789355385383 | Machine Learning | Himanshu Mittal |
| 46 | 9789355388414 | Manufacturing Automation | Chitragupt Swaroop Chitransh |
| 47 | 9789355387233 | Manufacturing Processes | Dheerendra Kumar Dwivedi |
| 48 | 9789355389886 | Materials Engineering | Dheerendra Kumar Dwivedi |
| 49 | 9789391505172 | Mathematics - I Calculus and Linear Algebra [For Non-Computer Science Engineering Branches] | Reena Garg |
| 50 | 9789391505271 | Mathematics - I Calculus and Linear Algebra [For Computer Science.Engineering Branches] | Reena Garg |
| 51 | 9789355384607 | Measurements & Metrology | V.P. Chandramohan |
| 52 | 9789355383730 | Mechanics of Materials | Arvind Vyavahare |
| 53 | 9789355383358 | Mechatronics, Robotics & Control | J. Srinivas |
| 54 | 9789355385116 | Microprocessor and Microcontroller | Saurabh Chaudhury, Risha Mal |
| 55 | 9789355386083 | Microprocessors | Yogendra Kumar Gupta |
| 56 | 9789355389169 | Mobile Communication and Networks | Vimal Bhatia |
| 57 | 9789355385475 | Operating Systems | Sukomal Pal |
| 58 | 9789355385604 | Operating Systems | Sanjaya Kumar Panda |
| 59 | 9789391505134 | Physics ( Oscillations, Waves and Optics) With lab manual | A.B. Bhattacharya, Atanu Nag |
| 60 | 9789391505257 | Physics ( Quantum Mechanics for Engineers) With lab manual | A.B. Bhattacharya, Atanu Nag |
| 61 | 9789391505165 | Physics (Introduction to Electromagnetic Theory) (with Lab Manual) | A.B. Bhattacharya, Atanu Nag |
| 62 | 9789391505059 | Physics (Introduction to Mechanics) (with Lab Manual) | A.B. Bhattacharya, Atanu Nag |
| 63 | 9789355385956 | Power Electronics: Theory and Practicals | Lalit Chandra Saikia |
| 64 | 9789355384942 | Power Systems-I | Suresh Mikkili |
| 65 | 9789355386236 | Power Systems-II | Arun Rahul S, Manas K. Jena |
| 66 | 9789355383754 | Product Innovation & Entrepreneurship | Vigneswaran C |
| 67 | 9789355384584 | Production & Operation Management | Navin Kumar Dev |
| 68 | 9789391505219 | Programming for Problem Solving (with Lab Manual) | R.S. Salaria |
| 69 | 9789355385451 | Signals and Systems | Sanjay L. Nalbalwar |
| 70 | 9789355386021 | Strength of Materials | Uday Shanker Dixit, Nelson Muthu, S. M. Kamal |
| 71 | 9789355383792 | Structural Engineering | T.D. Gunneswara Rao, D. Ravi Prasad |
| 72 | 9789355387004 | Surveying and Geomatics | P. K. Garg |
| 73 | 9789355384317 | Testing & Evaluation of Civil Engineering Materials | Sandeep Chaudhary, Kashyap A Patel |
| 74 | 9789355382306 | Theory of Computation | Debi Prasanna Acharjya |
| 75 | 9789355384928 | Transportation Engineering | A.K. Raji |
| 76 | 9789355385208 | VLSI Design | Mahendra Sakare, Shonal Chouksey |
| 77 | 9789391505332 | Workshop / Manufacturing Practices (with Lab Manual) | Veeranna D.K. |

## DIPLOMA COURSE BOOKS

| S.No. | ISBN | Book Name | Author |
|---|---|---|---|
| 1 | 9789355382481 | Aanandam An Act of Community Service | Shuchi Sharma |
| 2 | 9789355385567 | Advanced Manufacturing Processes | Ranjeet Kumar Sahu, Devendra Laxman Kamble |
| 3 | 9789355385109 | Advanced Surveying: Theory & Practice | Ramakant Agrawal, Parshottam Sarathe |
| 4 | 9789355384546 | Algorithms | Piyoosh P, Arnab Sarkar |
| 5 | 9789391505448 | Applied Chemistry (with Lab Manual) | Anju Rawlley, Devdatta Vinayakrao Saraf |
| 6 | 9789391505431 | Applied Physics-I (with Lab Manual) | Vinod Kumar Yadav, Mina Talati |
| 7 | 9789391505578 | Applied Physics-II (with Lab Manual) | Hussain Jeevakhan |
| 8 | 9789355387530 | Basic Mechanical Engineering | Dharmendra Singh, Chandra Shekhar Rajoria |
| 9 | 9789355387394 | Building Construction | Rinku Kumar, Sandeep Panchal |
| 10 | 9789355383280 | Building Electrification | Arnab Ghosh |
| 11 | 9789355384379 | Building Planning and Drawing | Aparna Dwivedi |
| 12 | 9789391505493 | Communication Skills in English | Anjana Tiwari |
| 13 | 9789355383570 | Computer Aided Machine Drawing Practice | Kanak Kalita |
| 14 | 9789355388964 | Computer Networking and Data Communication | Sanjaya Shankar Tripathy |
| 15 | 9789355382986 | Computer Networks: Theory & Practicals | Brijendra Pratap Singh, Manoj Madhava Gore |
| 16 | 9789355382672 | Computer Programming: Theory and Practicals | Satyendra Singh Chouhan |
| 17 | 9789355387714 | Computer System Organization | Sonal Yadav |
| 18 | 9789355382658 | Concrete Technology: Theory and Practice | Sasankasekhar Mandal, Vishwajit Anand, Sushil K. Agarwala |
| 19 | 9789355383334 | Construction Materials and Testing | Vanita Aggarwal |
| 20 | 9789355383501 | Consumer Electronics | Amit M. Joshi, Maulin M. Joshi, Urvashi Prakash Shukla |
| 21 | 9789355387684 | Data Structures: Theory & Practicals | NB Venkateswarlu |
| 22 | 9789355388513 | Design of Machine Elements | A. Kumaravel, M. Kathirselvam |
| 23 | 9789355388360 | Design of Steel and RCC Structure | Sparsh Johari |
| 24 | 9789355383303 | Digital Communication | Muralidhar Kulkarni, K. S. Shivaprakasha |
| 25 | 9789355387295 | Digital Electronics | Menka Yadav |
| 26 | 9789355389435 | Electric Circuits and Networks | Jamuna K., Nilanjan Tewari |
| 27 | 9789355388452 | Electric Motors and Transformers (Theory and Practicals) | Sanjay B. Bodkhe |
| 28 | 9789355385901 | Electric Power Transmission and Distribution | Sudhir Sharma |
| 29 | 9789355389466 | Electrical & Electronics Measurements and Instrumentation | Sudarsan Sahoo, Vipin Chandra Pal, Sudipta Chakraborty |
| 30 | 9789355382696 | Electronic Communication | M. D. Selvaraj, P. Nagaradjane |
| 31 | 9789355387202 | Electronic Devices and Circuits | Ankesh Jain |
| 32 | 9789355386373 | Electronic Equipment Maintenance | Chanchal Sharma |
| 33 | 9789355384690 | Embedded Systems | Deepa M, G. Santhanamari, Susithra. N |
| 34 | 9789355384867 | Energy Conservation and Audit | M. Rizwan, Majid Jamil |
| 35 | 9789391505509 | Engineering Graphics | Sharad K. Pradhan, K.K. Jain |
| 36 | 9789391505608 | Engineering Mechanics (with Lab Manual) | Bhankhar Bharat, Gokaldas Vandana Somkuwar |

## DIPLOMA COURSE BOOKS

| S.No. | ISBN | Book Name | Author |
|---|---|---|---|
| 37 | 9789391505516 | Engineering Workshop Practice | A. K. Sarathe |
| 38 | 9789391505653 | Environmental Science | Subrat Roy |
| 39 | 9789355384966 | Estimating, Costing and valuation | Sandeep Panchal |
| 40 | 9789355389510 | Fluid Mechanics & Hydraulic Machinery | Suman Chakraborty, Sourav Mitra, Aditya Bandopadhyay |
| 41 | 9789391505592 | Fundamentals of Electrical and Electronics Engineering (with Lab Manual) | Susan S. Mathew, Saji T. Chacko |
| 42 | 9789355387240 | Fundamentals of Power Electronics | Soumitra Kumar Mandal |
| 43 | 9789355389602 | Geotechnical Engineering (Theory & Practicals) | Neelima Satyam |
| 44 | 9789355386137 | Induction, Synchronous, and Special Electrical Machines | K. Selvajyothi, S. Tamilselvi, Sarat Kumar Sahoo |
| 45 | 9789355382290 | Industrial Engineering & Management | A. Manimaran |
| 46 | 9789355384959 | Internet of Things | Sujata Pal |
| 47 | 9789355388650 | Introduction to DBMS: Theory & Practicals | Myneni Madhu Bala |
| 48 | 9789355384294 | Introduction to e-Governance | Kamal Kant Verma |
| 49 | 9789355386564 | Introduction to Electric Generation Systems | Lalit Chandra Saikia, Nalin Behari, Deb Choudhury |
| 50 | 9789391505585 | Introduction to IT Systems (with Lab Manual) | Prashant Joshi |
| 51 | 9789355382900 | Linear Integrated Circuit | B. T. Krishna |
| 52 | 9789355388148 | Manufacturing Engineering | Santosh Kumar |
| 53 | 9789355383877 | Material Science & Engineering | Vinod Kumar, Nisheeth k. Prasad |
| 54 | 9789391505424 | Mathematics-I | Deepak Singh |
| 55 | 9789391505523 | Mathematics-II | Garima Singh |
| 56 | 9789355385994 | Measurements and Metrology | A. Gnanavelbabu, M. Santhana kumar, R. Saranraj |
| 57 | 9789355388131 | Mechanics of Materials and their Testing | R. Velmurugan, Pragathi |
| 58 | 9789355385048 | Microcontroller Applications | Ritula Thakur |
| 59 | 9789355385062 | Microcontrollers and Applications | Santanu Chattopadhyay |
| 60 | 9789355388100 | Mobile and Wireless Communication | Maulin Joshi, Urvashi P. Shukla |
| 61 | 9789355387790 | Production & Operations Management | Mayank Kumar Agrawal |
| 62 | 9789355388704 | Public Health Engineering | Rinku Kumar |
| 63 | 9789355387455 | Python Programming | Rupesh Nasre |
| 64 | 9789355387189 | Renewable Energy Technologies | Sanjay Agrawal, Rajeev K. Mishra |
| 65 | 9789355387479 | Software Engineering | Salman Abdul Moiz |
| 66 | 9789355389749 | Theory of Machines & Mechanisms | G. C. Mohan Kumar |
| 67 | 9789355387332 | Theory of Structures | Arunachalam Subramanian Balu |
| 68 | 9789355389015 | Thermal Engineering - I | V. Raghavan |
| 69 | 9789355389831 | Thermal Engineering - II | Parvesh Antil |
| 70 | 9789355386779 | Transportation Engineering (Theory & Practice) | A.K. Raji, K.K. Babu |
| 71 | 9789355384287 | Water Resource Engineering (Theory & Practice) | Prakash C. Swain, Dillip K. Ghose |
| 72 | 9789355383051 | Web Technologies: Theory & Practical | Raman Dugyala |

प्रो. टी. जी. सीताराम
अध्यक्ष
Prof. T. G. Sitharam
Chairman

सत्यमेव जयते

आज़ादी का अमृत महोत्सव

अखिल भारतीय तकनीकी शिक्षा परिषद्
(भारत सरकार का एक सांविधिक निकाय)
(शिक्षा मंत्रालय, भारत सरकार)
नेल्सन मंडेला मार्ग, वसंत कुंज, नई दिल्ली–110070
दूरभाष : 011–26131498
ई–मेल : chairman@aicte-india.org

ALL INDIA COUNCIL FOR TECHNICAL EDUCATION
(A STATUTORY BODY OF THE GOVT. OF INDIA)
(Ministry of Education, Govt. of India)
Nelson Mandela Marg, Vasant Kunj, New Delhi-110070
Phone : 011-26131498
E-mail : chairman@aicte-india.org

## FOREWORD

Engineers are the backbone of any modern society. They are the ones responsible for the marvels as well as the improved quality of life across the world. Engineers have driven humanity towards greater heights in a more evolved and unprecedented manner.

The All India Council for Technical Education (AICTE), have spared no efforts towards the strengthening of the technical education in the country. AICTE is always committed towards promoting quality Technical Education to make India a modern developed nation emphasizing on the overall welfare of mankind.

An array of initiatives has been taken by AICTE in last decade which have been accelerated now by the National Education Policy (NEP) 2020. The implementation of NEP under the visionary leadership of Hon'ble Prime Minister of India envisages the provision for education in regional languages to all, thereby ensuring that every graduate becomes competent enough and is in a position to contribute towards the national growth and development through innovation & entrepreneurship.

One of the spheres where AICTE had been relentlessly working since past couple of years is providing high quality original technical contents at Under Graduate & Diploma level prepared and translated by eminent educators in various Indian languages to its aspirants. For students pursuing 2nd year of their Engineering education, AICTE has identified 87 books, which shall be translated into 12 Indian languages - Hindi, Tamil, Gujarati, Odia, Bengali, Kannada, Urdu, Punjabi, Telugu, Marathi, Assamese & Malayalam. In addition to the English medium, books in different Indian Languages are going to support the students to understand the concepts in their respective mother tongue.

On behalf of AICTE, I express sincere gratitude to all distinguished authors, reviewers and translators from the renowned institutions of high repute for their admirable contribution in a record span of time.

AICTE is confident that these outcomes based original contents shall help aspirants to master the subject with comprehension and greater ease.

T. G. Sitharam
(Prof. T. G. Sitharam)

## ACKNOWLEDGEMENT

The authors are grateful to the authorities of AICTE, particularly Prof. T. G. Sitharam, Chairman; Prof. M. P. Poonia, Vice-Chairman; Prof. Rajive Kumar, Member-Secretary and Dr. Amit Kumar Srivastava, Director, Faculty Development Cell for their planning to publish the books on Consumer Electronics. We sincerely acknowledge the valuable contributions of the reviewer of the book Dr. Savitesh Madhulika Sharma, Professor and Dean (R&D), Siddartha Institute of Science and Technology Puttur (A.P) for making it students' friendly and giving a better shape in an artistic manner.

This book is an outcome of various suggestions of AICTE members, experts and authors who shared their opinion and thought to further develop engineering education in our country. Acknowledgements are due to the contributors and different workers in this field whose published books, review articles, papers, photographs, footnotes, references and other valuable information enriched us at the time of writing the book.

**Dr. Amit M. Joshi**

**Dr. Maulin M. Joshi**

**Dr. Urvashi Prakash Shukla**

# ACKNOWLEDGEMENT

The authors are grateful to the authorities of AICTE, particularly Prof. T. G. Sitharam, Chairman; Prof. M. P. Poonia, Vice-Chairman; Prof. Rajive Kumar, Member-Secretary and Dr. Amit Kumar Srivastava, Director, Faculty Development Cell for their planning to publish the books on Consumer Electronics. We sincerely acknowledge the valuable contributions of the reviewer of the book Dr. Savitesh Madhulika Sharma, Professor and Dean (R&D), Siddartha Institute of Science and Technology, Puttur (A.P.) for making it students' friendly and giving a better shape in an artistic manner.

This book is an outcome of various suggestions of AICTE members, experts and authors who shared their opinion and thought to further develop engineering education in our country. Acknowledgements are due to the contributors and different workers in this field whose published books, review articles, papers, photographs, footnotes, references and other valuable information enriched us at the time of writing the book.

Dr. Arun M. Joshi

Dr. Manish M. Joshi

Dr. Urvashi Prakash Shukla

# PREFACE

Reflecting past experience in classrooms, this text book on Consumer Electronics is written with a view to provide foundation of the subject. Using simple and lucid language, this text book aims to take reader for a journey to the important technical concepts of consumer electronics devices being used in daily life and offices. This book is suitable for use as one-semester course material for diploma students of Electronics, Electronics and communication engineering, Electrical and electronics engineering, Electronics & Instrumentation engineering.

The first edition of this book is organized in 6 modules. The outline of the book is as follows:

First unit describes the fundamentals of sound and its characteristics. Then after, a microphone's basic operating principle and its defining features are outlined. In addition, the differences and similarities between different types of microphones like carbon, moving coil, and cordless microphones have been discussed. Subsequently it explains how speakers work and their categorization. Unit also details three main types of sound recording—magnetic, direct, and optical.

Second unit of the book discusses fundamentals of a CD player, a sound system for a home theatre, a digital audio console, surround sound, and a public address (PA) system. It has also detailed of digital audio console's inner workings along with their various applications. In order to provide a deeper understanding, explanation as well as the functions for each component of a public address system is provided. The process to be followed to establish a public address system in different scenarios have been outlined in the module.

Television is the important part of communication and Fundamentals of television systems and its foundations of monochrome and colour television operations in the third unit of the book. It explains the standards used in monochrome television, including the scanning process and video signal components. Furthermore, the application of colour television is described in detail. This describes the additive and subtractive mixing techniques used to create various colours. In order for television programmes to be received by the receiver, they must comply with specific transmission standards, which are discussed in detail. It also discusses the various TV cameras and transmission of TV channels.

Knowledge of current generation television receiver systems and general technical know-how of sound and video systems are important for engineers. Fourth unit describes how the PAL-D colour TV receiver functions in depth. The television receiver includes a video amplifier, sound section, sync separator, and processor that have been thoroughly discussed. In addition, various types of video cables and interfaces have been discussed. Finally, the unit incorporates fundamentals of CD and DVD players. Types and functionality of various TVs namely Digital, LCD, LED, Plasma, HDTV, 3-D TV, and Projection TV have been briefly described, as well as their benefits and drawbacks stated. Additionally, the unit demonstrates the operation of the outdoor and indoor units of direct-to-home (DTH) technology.

Microwave oven and the washing machine, two staples of the modern household. Microwave oven basic operating principle, components, technical specifications and necessary user precautions have also been outlined in the fifth unit. In the second part, various types of washing machines have been broken down into their component parts and discussed in greater detail. In addition, the controller, fuzzy logic, and technical specifications of a washing machine have all been discussed

Various digital electronic devices including a fax machine, photocopier, air conditioner, refrigerator, digital camera, and camcorder are frequently found in daily affairs. Understanding of basic operational principles and fault diagnosis is required in order to enhance the life span of these products. Sixth module of the book delves deeply into the fax machine's details, its origins, functionalities, operation, and the basics of sending and receiving faxes. Additionally, the components and operating principles of the photocopier, also known as electrophotography, have been covered. One of the most widespread types of technology, air conditioners come in a wide variety of models. Moreover, aspects of AC operation are also discussed in this unit. A number of different parts, including the refrigerant fluid, compressor, condenser, thermostatic expansion valves, evaporator, and receiver, carry out a refrigerator's cooling process. Digital cameras, including their operation and classification, have also been outlined. Camcorder fundamentals have been highlighted in the last section of the chapter. Besides giving several of multiple choice questions as well as questions of short and long answer types marked in two categories following lower and higher order of Bloom's taxonomy, assignments through a number of numerical problems, a list of references and suggested readings are given in the unit so that one can go through them for practice.

After the related practical, based on the content, there is a "Know More" section. This section has been carefully designed so that the supplementary information provided in this part becomes beneficial for the users of the book. This section mainly highlights the initial activity, examples of some interesting facts, analogy, history of the development of the subject focusing the salient observations and finding, timelines starting from the development of the concerned topics up to the recent time, applications of the subject matter for our day-to-day real life or/and industrial applications on variety of aspects, case study related to environmental, sustainability, social and ethical issues whichever applicable, and finally inquisitiveness and curiosity topics of the unit.

**Dr. Amit M. Joshi**
**Dr. Maulin M. Joshi**
**Dr. Urvashi Prakash Shukla**

# OUTCOME BASED EDUCATION

For the implementation of an outcome-based education the first requirement is to develop an outcome-based curriculum and incorporate an outcome-based assessment in the education system. By going through outcome-based assessments, evaluators will be able to evaluate whether the students have achieved the outlined standard, specific and measurable outcomes. With the proper incorporation of outcome-based education there will be a definite commitment to achieve a minimum standard for all learners without giving up at any level. At the end of the programme running with the aid of outcome-based education, a student will be able to arrive at the following outcomes:

Programme Outcomes (POs) are statements that describe what students are expected to know and be able to do upon graduating from the program. These relate to the skills, knowledge, analytical ability attitude and behavior that students acquire through the program. The POs essentially indicate what the students can do from subject-wise knowledge acquired by them during the program. As such, POs define the professional profile of an engineering diploma graduate.

National Board of Accreditation (NBA) has defined the following seven POs for an Engineering diploma graduate:

**PO1. Basic and Discipline specific knowledge:** Apply knowledge of basic mathematics, science and engineering fundamentals and engineering specialization to solve the engineering problems.

**PO2. Problem analysis:** Identify and analyses well-defined engineering problems using codified standard methods.

**PO3. Design/ development of solutions:** Design solutions for well-defined technical problems and assist with the design of systems components or processes to meet specified needs.

**PO4. Engineering Tools, Experimentation and Testing:** Apply modern engineering tools and appropriate technique to conduct standard tests and measurements.

**PO5. Engineering practices for society, sustainability and environment:** Apply appropriate technology in context of society, sustainability, environment and ethical practices.

**PO6. Project Management:** Use engineering management principles individually, as a team member or a leader to manage projects and effectively communicate about well-defined engineering activities.

**PO7. Life-long learning:** Ability to analyse individual needs and engage in updating in the context of technological changes.

## COURSE OUTCOMES

By the end of the course the students are expected to learn:

**CO-1:** Understand and Compare Types of Microphones and Speakers.

**CO-2:** Maintain and Troubleshoot the Audio Systems

**CO-3:** Understand the Operating Principle of Color TV

**CO-4:** Understand the Architecture of TV Receiver System

**CO-5:** Maintain and troubleshoot Electronic Appliances

**CO-6:** Understand the Electromechanical systems used in Electronic Appliances

Mapping of Course Outcomes with Programme Outcomes to be done according to the matrix given below:

| Course Outcomes | Expected Mapping with Programme Outcomes (1- Weak Correlation; 2- Medium correlation; 3- Strong Correlation) | | | | | | |
|---|---|---|---|---|---|---|---|
| | PO-1 | PO-2 | PO-3 | PO-4 | PO-5 | PO-6 | PO-7 |
| CO-1 | 3 | 3 | 3 | 3 | 1 | 1 | 3 |
| CO-2 | 3 | 2 | 2 | 2 | 1 | 1 | 3 |
| CO-3 | 3 | 2 | 2 | 2 | 1 | 1 | 3 |
| CO-4 | 3 | 2 | 2 | 3 | 1 | 1 | 3 |
| CO-5 | 3 | 3 | 3 | 3 | 1 | 1 | 3 |

## GUIDELINES FOR TEACHERS

To implement Outcome Based Education (OBE) the knowledge level and skill set of the students should be enhanced. Teachers should take a major responsibility for the proper implementation of OBE. Some of the responsibilities (not limited to) for the teachers in OBE system may be as follows:

- Within reasonable constraints, they should manoeuvre time to the best advantage of all students.
- They should assess the students only upon certain defined criterion without considering any other potential ineligibility to discriminate against them.
- They should try to grow the learning abilities of the students to a certain level before they leave the institute.
- They should try to ensure that all the students are equipped with quality knowledge as well as competence after they finish their education.
- They should always encourage the students to develop their ultimate performance capabilities.
- They should facilitate and encourage group work and team work to consolidate newer approaches.
- They should follow Bloom's taxonomy in every part of the assessment.

**Bloom's Taxonomy**

| Level | Teacher should Check | Student should be able to | Possible Mode of Assessment |
|---|---|---|---|
| Create | Students ability to create | Design or Create | Mini project |
| Evaluate | Students ability to justify | Argue or Defend | Assignment |
| Analyse | Students ability to distinguish | Differentiate or Distinguish | Project/Lab Methodology |
| Apply | Students ability to use information | Operate or Demonstrate | Technical Presentation/ Demonstration |
| Understand | Students ability to explain the ideas | Explain or Classify | Presentation/Seminar |
| Remember | Students ability to recall (or remember) | Define or Recall | Quiz |

## GUIDELINES FOR STUDENTS

Students should take equal responsibility for implementing the OBE. Some of the responsibilities (not limited to) for the students in OBE system are as follows:

- Students should be well aware of each UO before the start of a unit in each and every course.
- Students should be well aware of each CO before the start of the course.
- Students should be well aware of each PO before the start of the programme.
- Students should think critically and reasonably with proper reflection and action.
- Learning of the students should be connected and integrated with practical and real life consequences.
- Students should be well aware of their competency at every level of OBE.

# ABBREVIATIONS AND SYMBOLS

## List of Abbreviations

| General Terms | | | |
|---|---|---|---|
| **Abbreviations** | **Full form** | **Abbreviations** | **Full form** |
| A/C | Air Conditioning | DVD | Digital Versatile Disc |
| AC | Alternating Current | DVI | Digital Visual Interface |
| ADC | Analogue-to-Digital Converter | EFM | Eight-to-Fourteen Modulation |
| AFT | Automatic Frequency tuning | EMF | Electro Motive Force |
| AHU | Air Handling Unit | EIA | Electronic Industries Alliance |
| AI | Artificial Intelligence | Fax | Facsimile |
| AGC | Automatic Gain Control | FCU | Fan Coil Unit |
| AM | Amplitude modulation | FPS | Frames Per Second |
| APC | Automatic Phase Control | FM | Frequency Modulation |
| CCFL | Cold Cathode Fluorescent Lamps | H | Height |
| C.C.I.R. | Consultative Committee for International Radio | HD | High-Definition |
| CCD | Charge Coupled Device | HDMI | High-Definition Multimedia Interface |
| CCTV | Closed Circuit Television | HDTV | High Definition TV |
| CD | Compact Disc | HVAC system | Heating, Ventilation, and Air conditioning system |
| CEA | Central Electricity Authority | I/O | Input/Output |
| CMOS | Complementary Metal–Oxide–Semiconductor | IC | Integrated Circuit |
| CPU | Central Processing Unit | IF | Intermediate Frequency |
| CRT | Cathode-Ray Tube | ISO | International Standards Organization |
| CVD | Composite Video | LED | Light-Emitting Diode |
| DAC | Digital-to-Analog Converter | LCD | Liquid Crystal Display |
| DBS | Direct Broadcast Satellite | LNA | Low Noise Amplifier |
| DC | Digital Current | MPEG | Moving Picture Experts Group |
| DDWG | Digital Display Working Group | NTSC | National Television Standards Committee |
| DSLR | Digital Single-Lens Reflex | OLED | Organic Light-Emitting Diode |
| DSB | Double Sideband | OTT | Over The Top |
| DTH | Direct-to-Home | PAL | Phase Alternate Line |
| DTV | Digital Television | PAL-D | Phase Alternating Line- Delay |
| RAM | Random-Access Memory | TLR | Twin-Lens Reflex |
| RCA | Radio Corporation of America | TV | Tele Vision |

| General Terms | | | |
|---|---|---|---|
| **Abbreviations** | **Full form** | **Abbreviations** | **Full form** |
| RF | Radio Frequency | UHF | Ultra-High Frequency |
| RMD | Resist Mater Disc | UV | Ultra Violet |
| S-Video | Separate Video | UXGA | Ultra-Extended Graphics Array |
| SDTI | Serial Data Transport Interface | VCR | Video Cassette Recorder |
| SDI | Serial Digital Interface | VHF | Very High Frequency |
| SNR | Signal-to-Noise Ratio | VHS | Video Home System |
| SLR | Single-Lens Reflex | VHS-C | Video Home System – Compact |
| SPL | Sound Pressure Levels | VSB | Vestigial Side Band |
| Sync | Synchronization | 2D | Two Dimensional |
| TFT | Thin Film Transistor | 3D | Three Dimensional |

## List of Symbols

| Symbols | Description | Symbols | Description |
|---|---|---|---|
| cm | Centi-meter | mm | Milli-meter |
| dB | Decibels | ms | Milli-seconds |
| GHz | Giga Hertz | $N/m^2$ | Newtons per square meter |
| Hz | Hertz | Ω | Ohms |
| kHz | Kilo Hertz | Pa | Pascal |
| kW | Kilo Watt | s | Seconds |
| kg | Kilo-gram | V | Volt |
| MHz | Mega Hertz | W | Watt |
| m | Meters | $W/m^2$ | Watts per square meter |
| μs | Micro seconds | | |

# LIST OF FIGURES

# LIST OF TABLES

# CONTENTS

# 1 Audio Fundamentals and Devices

## UNIT SPECIFICS

Through this unit, we have discussed the following aspects:

- Understand various types of Microphones and Speakers
- Troubleshoot the Audio Systems
- Study of basic Operating Principle of Color TV
- Analysis of TV Receiver System
- The basic troubleshoot of various Electronic Appliances
- Understand the Electromechanical systems used in Electronic Appliances

The topics are discussed with various examples for generating further curiosity and creativity and improving problem-solving capacity.

Besides giving a large number of multiple-choice questions as well as questions of short and long answer types marked in two categories following lower and higher order of Bloom's taxonomy, assignments through several numerical problems, a list of references, and suggested readings are given in the unit so that one can go through them for practice.

After the related practical, based on the content, there is a "Know More" section. This section has been carefully designed so that the supplementary information provided in this part becomes beneficial for the users of the book. This section mainly highlights the initial activity, examples of some interesting facts, analogy, history of the development of the subject focusing on the salient observations and finding, timelines starting from the development of the concerned topics up to the recent time, applications of the subject matter for our day-to-day real life or/and industrial applications on a variety of aspects, case study related to environmental, sustainability, social and ethical issues whichever applicable, and finally inquisitiveness and curiosity topics of the unit.

## RATIONALE

This unit describes the fundamentals of sound, including its loudness and volume, pitch, frequency response, fidelity, sensitivity, and selectivity, as measured by audio level metres. Then after, a microphone's basic operating principle and its defining features are outlined. In addition, the differences and similarities between carbon, moving coil, and cordless microphones have been discussed. This module not only explains how speakers work and what qualities they should have but also categorised them. The three main types of sound recording—magnetic, direct, and optical—have been outlined with suitable diagrams.

## PRE-REQUISITE

Mathematics: Electronics Circuits (Class XII)

Physics: Signal and Wave (Class XII)

## UNIT OUTCOMES

Outcomes of this unit are: After learning this unit student will be able to

**U1-O1:** Explain the basic characteristics and parameters of the sound signal

**U1-O2:** Describe characteristics, working principles and types of microphone

**U1-O3:** Describe characteristics, working principles and types of speakers

**U1-O4:** Explain the type of sound recording

| Unit-1 Outcomes | EXPECTED MAPPING WITH COURSE OUTCOMES (1- Weak Correlation; 2- Medium Correlation; 3- Strong Correlation) | | | | | |
|---|---|---|---|---|---|---|
| | CO-1 | CO-2 | CO-3 | CO-4 | CO-5 | CO-6 |
| U1-O1 | 2 | 3 | 2 | 1 | 2 | 2 |
| U1-O2 | 3 | 3 | 1 | 1 | 1 | 1 |
| U1-O3 | 3 | 3 | 2 | 2 | 2 | 1 |
| U1-O4 | 2 | 2 | 2 | 1 | 1 | 1 |

## 1.1 BASIC CHARACTERISTICS OF SOUND SIGNAL

Sound waves are longitudinal propagation of consecutive compression and rarefaction travel through a medium. As a sound wave hits your eardrum, it is transformed into an electrical signal, which is then transmitted by your auditory nerves to your brain and is interpreted as sound. It exhibits the full spectrum of wave properties, including amplitude, frequency, velocity, wavelength, and phase. Pure tonal intensity is depicted in (Fig. 1.1) as a sine wave.

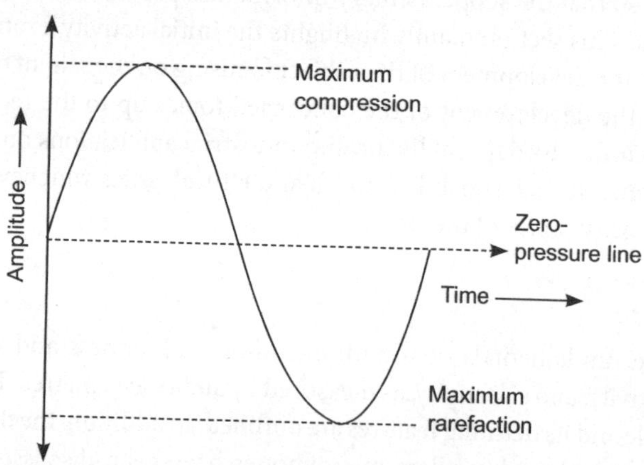


**Fig. 1.1:** Single cycle of pressure variation represented as a sine wave

The parameters by which it operates are specified below.

(a) **Amplitude:** It determines the strength of compressions and rarefactions in a medium.

(b) **Frequency (f):** It represents the rate at which compressions and rarefactions alternate over one second time period. Sound waves can range from 20 to 20,000 hertz (Hz).

(c) **Time period (T):** It is the amount of time, in seconds, required for one full cycle of oscillation.

$$T = \frac{1}{f} \quad ...(1.1)$$

(d) ***Velocity:*** Is the distance travelled in one second and it is proportional to the change in temperature.

$$v_2 = v_1 \times \sqrt{\frac{t_2}{t_1}} \quad ...(1.2)$$

where, $v_1$ = velocity at °K $t_1$ and $v_2$ = velocity at $t_2$ °K

(e) ***Wavelength:*** Is a measure of how far apart the maximum compression is.

$$\lambda = \frac{v}{f} \quad ...(1.3)$$

### 1.1.1 Audio Level Metering

When a sound wave travels through a medium, it causes pressure changes, which can be either compression or rarefaction. The higher the decibel level, the more compression and rarefaction will occur. The amplitude of a sound wave is measured in Pascal's (Pα) or Newton per square metre $\left(\frac{N}{m^2}\right)$ of pressure. Sound wave intensity is measured in terms of the average rate of energy flow through a one square metre cross section at right angles to the direction of motion. Watts per square metre $\left(\frac{w}{m^2}\right)$ is the unit of measurement for the intensity of sound energy. Consequently, it is proportional to the square of the amplitude. Sound intensity levels are typically expressed in decibels (dB) relative to the human hearing threshold.

$$\text{Decibel (dB) for powers} = 10 \log \left(\frac{P_1}{P_2}\right) \quad ...(1.4)$$

$$\text{Decibel (dB) for amplitudes} = 20 \log \left(\frac{A_1}{A_2}\right) \quad ...(1.5)$$

where $P_1$ and $P_2$ are output and input power respectively and $A_1$ and $A_2$ are output and input voltage amplitude respectively.

### 1.1.2 Decibel Level in Acoustic Measurement

**Table 1.1:** Intensity of a wide range of sounds

| Sound type | Pressure (Pa or $\left(\frac{N}{m^2}\right)$) | Intensity $\left(\frac{W}{m^2}\right)$ | Intensity (dB): over the threshold of hearing |
|---|---|---|---|
| Hearing threshold | $20 \times 10^{-6}$ | $10^{-12}$ | 0 |
| Leaves rustling | $63 \times 10^{-6}$ | $10^{-11}$ | 10 |
| Whisper | $20 \times 10^{-5}$ | $10^{-10}$ | 20 |
| Average residence | $20 \times 10^{-4}$ | $10^{-8}$ | 40 |
| Violen's soft note at 3m | $31 \times 10^{-4}$ | $2.5 \times 10^{-8}$ | 44 |
| Typical conversations | $63 \times 10^{-4}$ | $10^{-7}$ | 50 |
| Average sound of an automobilec | $20 \times 10^{-3}$ | $10^{-6}$ | 60 |
| Normal speech | 0.1 | $0.25 \times 10^{-4}$ | 74 |
| Heavy traffic | 0.2 | $10^{-4}$ | 80 |
| Thunder | 2 | 10-2 | 100 |
| Loud band in a small hall | 20 | 1 | 120 |
| Threshold of pain | 63 | 10 | 130 |

### 1.1.3 LEVEL AND LOUDNESS

Sound intensity, also known as sound pressure, is the most accurate physical indicator of loudness when it is analysed near the eardrum. The frequency content, duration, and environment in which a sound is presented are just a few of the many other factors that can affect how loud it is. Higher amplitudes are linked to louder sounds, and the loudness of a specific sound is related to the amplitude of the sound wave. Moreover, a logarithmic unit of sound intensity, the decibel (dB), is used to quantify loudness. A food processor, a heavy truck (25 feet away), a power lawnmower, a subway train (20 feet away), and live rock music are among the sounds that could cause hearing damage at levels between 80 dB and 130 dB.

Sound pressure levels (SPL) are measured in decibels (dB) units. A decibel measures loudness in relation to another source of loudness. The SPL indicator employs a reference sound pressure ($P_o$) of 20 micro pascals. A tenfold increase in pressure wave amplitude corresponds to a doubling of SPL, or sound pressure level.

$$SPL = 20 * \log\left(\frac{P}{P_o}\right) \qquad ..(1.6)$$

Where P = sound pressure

### 1.1.4 PITCH

The term "pitch" is used to describe the audible quality of a sound based on its perceived frequency. Furthermore, when there are no harmonics (pure tone) present in the sound, frequency alone is used to identify the pitch. On the other hand, the frequency and intensity of a sound determine its pitch in speech and music. It can be observed in (Fig. 1.2) that higher-pitched sounds have a higher-frequency wave, while lower-pitched sounds have a lower-frequency wave.

The frequency 440 hertz (Hz) frequency is considered as pitch. The ear is unaffected by the relative phases of a complex wave's component tones when judging its quality or timbre. Therefore, it can be concluded that the human ear is resistant to phase shifts.

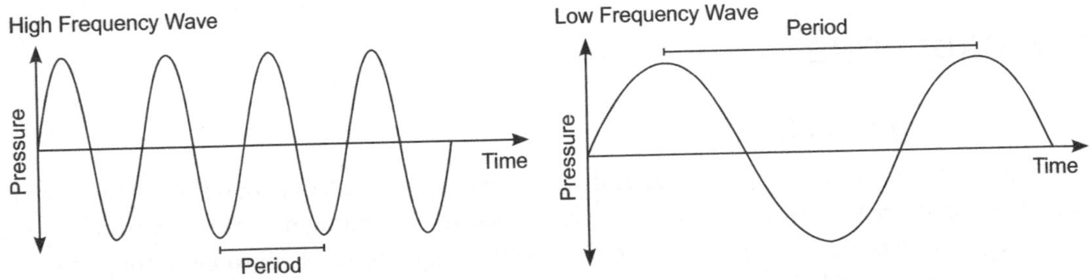


**Fig. 1.2:** Pressure v/s time graph of a sound wave.

## 1.1.5 FREQUENCY RESPONSE

The frequency response of an audio component is the frequency range it can accurately reproduce. Moreover, it indicates whether or not it alters the signal and how well it reproduces all audible frequencies. Therefore, we can define it as the range of frequencies within ± 1 dB of the microphone's 1000 Hz output.

Both electrical inductance and capacitance can be thought of as analogous to the mass of the vibrating system and its compliance. The effect of mass is attenuation at high frequencies and compliance at low frequencies. Because of its movable parts, a microphone has a resonant frequency, where the signal is amplified.

## 1.1.6 FIDELITY

The term "fidelity" is used to describe the degree to which an audio copy is faithful to the original. The term "high fidelity" (also spelt "Hi-Fi") first gained popularity in the 1950's to describe audio components and recordings that were able to faithfully recreate the original recording. The following are qualities that ought to characterise the ideal fidelity:

1. Infinite signal-to-noise ratio (SNR)
2. No frequency distortion
3. No amplitude distortion (non-linear distortion)
4. No spatial distortion
5. High dynamic range: 0 dB – 130 dB
6. The space in which the sound is being reproduced should be made to feel like its original environment.

## 1.1.7 SENSITIVITY AND SELECTIVITY

The human ear is extremely sensitive and able to detect sound with an intensity as low as 0.1 $\frac{pw^2}{m}$. In addition, the ear is sensitive not to absolute but to relative values of intensity (or dB). The selectivity of a receiver is defined as its capacity to filter out unwanted signals while retaining the ones that are of interest. The human ear possesses the following attributes:

1. For all ages, the ear is the most sensitive from 3 – 4 kHz.
2. As age increases, the sensitivity of the ear to high frequencies decreases.
3. For frequencies below 500 Hz, the sensitivity of the ear is low for all age groups.

## 1.2 MICROPHONE

### 1.2.1 Working Principle

Microphones are transducers that measure and record changes in ambient sound levels by means of minute electrical currents. When sound waves are applied to a diaphragm, it vibrates, which in turn causes a magnet to move in close proximity to a coil which may be flexible in some configurations. In contrast, condenser microphones use capacitance as their operating principle. Parallel conducting plates in a capacitor store charge and dampen fluctuations in a signal, such as those caused by a power supply's voltage. The incoming sound in a condenser microphone causes vibrations in one plate of a capacitor. In order to make sense of the varying capacitance, an electrical signal must be generated.

The following are the steps about how a microphone converts sound energy into electrical energy:

1. The energy of your voice travels in the form of sound waves that are emitted when you speak into a microphone.
2. The sound waves that strike a microphone cause the diaphragm inside to move back and forth. Consequently, the coil that is attached to the diaphragm also vibrates.
3. The magnetic field generated by the permanent magnet penetrates the coil and an electric current is generated in the coil as it oscillates within the magnetic field.
4. At last, the microphone discharges its electric current to an external sound amplifier or recorder.

### 1.2.2 Characteristics

There are a few key factors that determine a microphone's overall quality:

1. **Sensitivity:** A microphone's sensitivity is a quantitative indicator of how well it functions as a transducer. Its sensitivity is measured in terms of the ratio of the voltage it produces (the intensity of the audio signal) to the sound pressure level to which it is exposed. It is measured in millivolts per one pascal of pressure at one thousand hertz.
2. **Signal-to-noise ratio (SNR):** There is internal noise in a microphone due to the circuit's resistance, the transformer, and other components and the noise level is represented as the corresponding sound pressure level (SPL). Therefore, SNR is defined as the dB ratio of the output at 1 Pascal sound pressure level to the output at silence. The SNR is the comparison between the strength of the signal of interest and the strength of any interfering noise. A high SNR indicates a well-operating microphone.
3. **Frequency response:** In order to quantify the magnitude and phase of the output as a function of the input frequency, we need to measure the frequency response of the system. Components are designed to provide a consistent response across the system's frequency range to reduce audible distortion. A microphone's frequency response is the range of audible frequencies within ±1 dB of the output at 1000 Hertz.

4. **Distortion:** Microphones are prone to three types of distortion: frequency distortion, non-linear distortion, and phase distortion.
5. **Directivity:** Polar diagrams are commonly used to illustrate the directional sensitivity pattern of a microphone, which is known as its directivity. Half-power points on a polar diagram depict the microphone's directivity.
6. **Output impedance:** It is the type of complementary transformer needed to efficiently transfer power from the microphone to the transmission line and then to the amplifier is determined by its output impedance, measured in ohms.

## 1.2.3 TYPES OF MICROPHONES

**Table 1.2:** Analysing the differences between various microphone designs

| Parameter | Moving Coil | Ribbon | Crystal | Capacitor | Carbon |
|---|---|---|---|---|---|
| Sensitivity | Without transformer: 30μV With transformer: 90μV | Without transformer: 3μV With transformer: 90μV | 50 × 103 | 3 × 103 | 100 × 103 |
| Self-noise | Less than carbon | Less than moving coil | Less than ribbon | Lowest | Highest |
| Noise pick-up | High | Low | More than ribbon | Low | Lowest |
| Frequency response | 60-1000 Hz | 20-1200 Hz | 100-8000 Hz | 40-15000 Hz | 200-5000 Hz |
| Natural resonance frequency | 3-4 kHz | 20 Hz | 6000 Hz | 6000 Hz | 2000 Hz |
| Distortion | 5% | 1% | 1% | 1% | 10% |
| Basic directivity | Omni-directional | Bidirectional | Omni-directional | Omni-directional | Omni-directional |
| Output impedance (ohms) | 25 | 0.25 | 1 x 106 | 100 × 106 | 100 |
| Ruggedness | Rugged | Most delicate | Less than moving coil | Delicate | Most rugged |
| Effect of temperature | Minimal | Minimal | Yes | Minimal | Yes |
| Effect of moisture | No | No | Yes | No | Yes |
| Basic supply | Not required | Not required | Not required | Sometimes required | Required |

| Parameter | Moving Coil | Ribbon | Crystal | Capacitor | Carbon |
|---|---|---|---|---|---|
| Distance of speaker (cm) | 25 | 50 | Close | Close | Close |
| Size | Large | Large | Small | Large | Small |
| Cost | Average | High | Low | High | Lowest |
| Applications | PA systems, broadcast, etc. | Drama, music, etc. | Home recording system | Professional recording | Telephone |

### 1.2.4 CARBON MICROPHONE

Carbon microphones were primarily used in public address (PA) systems, broadcasting, military operations, and telephones. The rugged microphones had a high output level and were cheap to manufacture because of their simplistic design. It is a device composed of carbon granules encased in a container which is connected in series with a load through a DC power supply and the load current will fluctuate in response to changes in the carbon granules' pressure. These fine carbon granules are sandwiched between two metal plates, with the upper plate (the diaphragm) connected to a movable metal plate via a metal piston or plunger. Further, the metal plate below the diaphragm is permanently installed and separated from it. The device is shielded by a protective cover with holes.

When the load is connected to the metal plates, water passes through the carbon granules and the load from a battery connected between the metal plates. Furthermore, the output transformer's job is to filter any DC noise from the microphone's signal. (Fig 1.3) illustrates the construction of a typical carbon microphone.

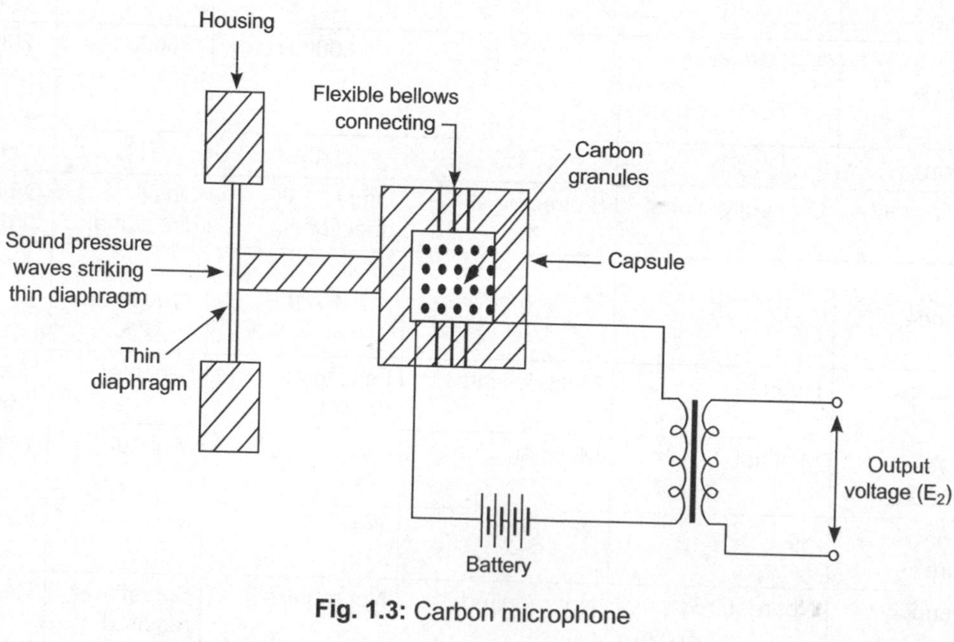


Fig. 1.3: Carbon microphone

### 1.2.5 MOVING COIL MICROPHONE

The principle of electromagnetic induction is used by the moving-coil (dynamic) microphone; when sound pressure variations move a coil within a magnetic field, the magnetic flux through the coil changes. As a result, the microphone's output is an EMF in the coil. The EMF's magnitude is determined by the flux's rate of change, and consequently, the coil's velocity in turn. The coil's displacement is determined by the force of the sound waves exerted on the diaphragm. Its primary parts are a magnet, a diaphragm, and a coil. Moreover, the equivalent electrical circuit of a moving-coil microphone is depicted in (Fig. 1.4). Its primary applications are in PA systems and broadcasts.

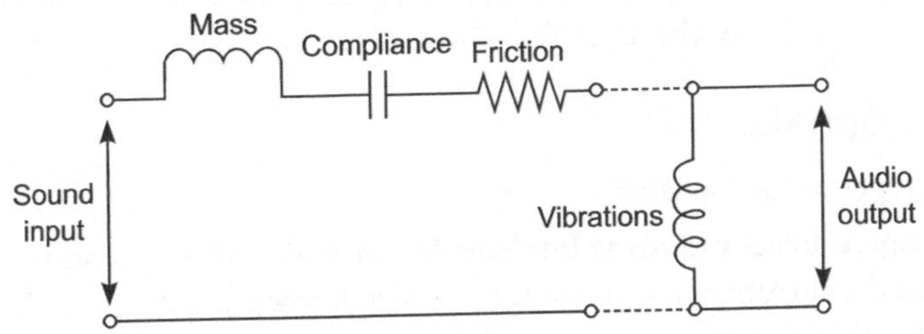


**Fig. 1.4** Equivalent electrical circuit of a moving-coil microphone.

### 1.2.6 CORDLESS MICROPHONE

A small frequency-modulated very high frequency (VHF) transmitter of low power is used in wireless microphones, also known as radio microphones. Merely a few mill watts are all that is being used here. Moreover, the microphone does not require a cable to connect it to the amplifier. A VHF receiver at an appropriate distance picks up the signal, amplifies it, and sends it to the speakers. It serves a purpose during sports oath-taking ceremonies.

## 1.3 SPEAKERS

### 1.3.1 Working Principle and Characteristics

A transducer that can effectively transform audio-frequency electrical signals into audible sound waves is called a loudspeaker. Moreover, this component is also known as output transducer. The following are the basic characteristics of the ideal speakers:

1. **Efficiency**: It is measured by how well the output sound levels measure upto the input sounds (electrical power). The value is determined by how closely the mechanical impedance matches the acoustic impedance of the disturbed air volume.
2. **Noise**: The unwanted sound that is not part of the input signal but is still heard after passing through a loudspeaker is referred to as "noise". More crucial is the system's SNR, which is the comparison of the signal output to the noise output.
3. **Frequency response**: It describes how well it reproduces sounds within the human hearing range. For the most accurate sound reproduction, a loudspeaker's response between 20 Hz and 20 kHz should be flat to within +1 dB.
4. **Distortion**: It refers to any alteration of the frequency, phase, or amplitude characteristics of the output sound relative to the input audio signal. The mass and compliance effect may cause

frequency and phase distortion. As the coil moves through a magnetic field, any inhomogeneities will cause distortions in the signal's amplitude or non-linearity.

5. **Directivity**: It is measured by comparing the sound pressure level at a given location (in the direction of maximum intensity) to the sound pressure level that would have been present at that location had the loudspeaker been omnidirectional.
6. **Power:** The power rating indicates the maximum amount of audio power (in watts) that it can handle. However, the speaker will be permanently damaged if you use more than the recommended amount of power.
7. **Impedance**: For a maximum output from the source amplifier, the loudspeaker's input impedance should be the same as that of the amplifier. Ohms (Ω) is used as the unit of measurement

## 1.3.2 Types of Speakers

Here are the various types of loudspeakers:

1. **Direct Radiating: A direct radiating loudspeaker operates using the same** interaction between a magnetic field and a current as an AC motor. It is also known as a loudspeaker with a moving cone. A coil is placed within a uniform magnetic field, and the audio signal in form of current interacts with the present magnetic field. This causes a force proportional to the audio current to act upon the movable coil. Inducing vibratory motion in the coil causes a conical paper diaphragm to vibrate and generate pressure vibrations in the air, resulting in the generation of sound waves. However, cone speakers have considerably low efficiency and are limited to mid-range frequencies. Typically, the SNR exceeds 30 dB. The magnetic flux density is non-uniform due to non-linearity, which may result in significant amplitude distortion.

Fig. 1.5: Direct radiating loudspeaker.[1]

2. **Indirect Radiating:** Indirect radiating loudspeakers use a moving coil placed in a magnetic field, but instead of radiating acoustic power directly in the open space of the listeners' area, the power is primarily delivered to the air trapped in a fixed, non-vibrating, tapered horn, and then to the

air in the listeners' area. Horn speaker functions as an acoustic transformer, that allows better impendence match between the air's low impedance and the voice coil assembly's high impedance. Consequently, increasing its efficiency. Moreover, a larger horn is required to enhance low frequency response.

**Fig. 1.6:** Indirect radiating loudspeaker.[2]

3. **Multi-Speaker System:** A line source speaker is when multiple drive units are stacked on top of one another in an appropriate enclosure, the loudspeaker from a line of source. Because they are

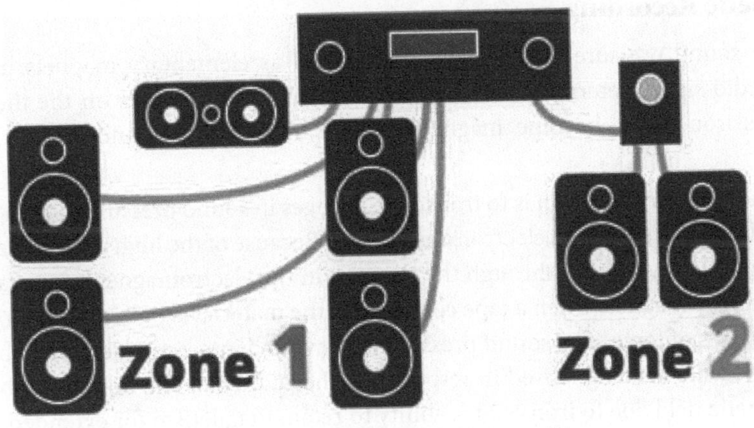


**Fig. 1.7:** Multi-speaker system.[3]

arranged in a column, it is also known as column speaker. (Fig 1.8) depicts five speakers arranged in a column. On the axis of the system, all the drive units' sound waves are in phase and, as a result, they reinforce each other in the front. Away from the axis, the path lengths of various loudspeakers will vary, resulting in phase differences and there will be a reduction in side sound due to the differences. Consequently, the column would emit the most sound in the front.

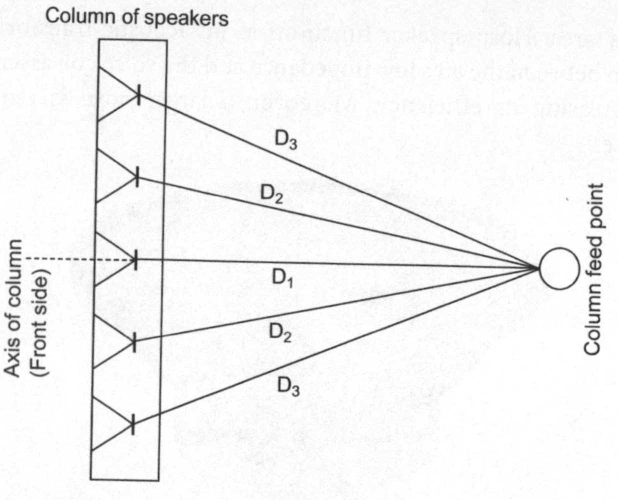


**Fig. 1.8** A column of loudspeakers fed by a central point

## 1.3.3 Sound Recording

### 1.3.3.1 Magnetic Recording

The variations in sound pressure are recorded and stored as elementary magnets using a magnetic recording. The audio signal determines the range and intensity. It operates on the theory that certain materials, such as iron oxide, become magnetised in a magnetic field and retain their magnetism permanently.

A microphone's primary function is to transduce changes in sound pressure into an electrical signal. This signal is amplified and sent to an electromagnet's coil. Because of the high reluctance of air, magnetic lines of force have difficulty passing through the tiny gap in the electromagnet known as the head. Iron oxide forms elementary magnets when a tape coated with the material is pulled across a gap, providing a direct path for the lines of force. The sound pressure is converted into a magnetic field by the iron oxide coating on the tape, which is magnetised in response to the audio current. Sound has been recorded as a fluctuating magnetic field due to iron oxide's ability to retain magnetism for extended periods of time.

The tape is fed back through a gap like the one it went through to record the sound, and this time the force lines in the coil shift to reproduce the original sound. This causes an electromagnetic field (audio signal) to be induced in the coil, the strength of which is proportional to the tape's rate of magnetic flux change. Thereafter, the amplified induced EMF is sent to a loudspeaker, which converts the signal into an audible sound.

### 1.3.3.2 Digital Recording

Sound recorded digitally offers nearly total immunity to noise, which led to the development of a CD system. To store its digital audio, a CD uses pits measuring 1μm deep and 0.5 μm wide, but their length can vary (1 μm – 3 μm). Furthermore, the track pitch (the distance between two parallel tracks) is 1.6 μm (Fig 1.9).

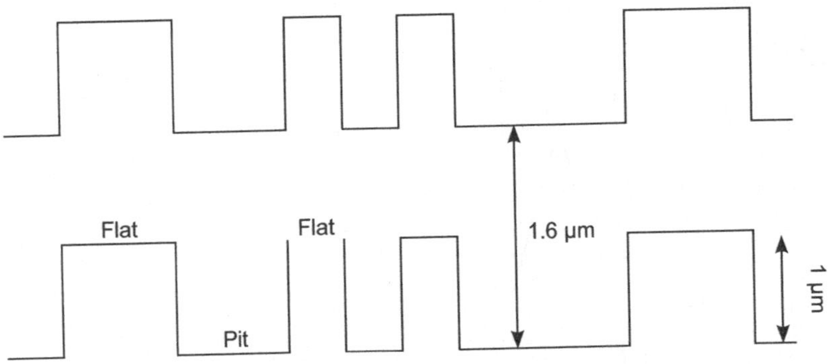


**Fig. 1.9:** Pits' dimensions on CD tracks

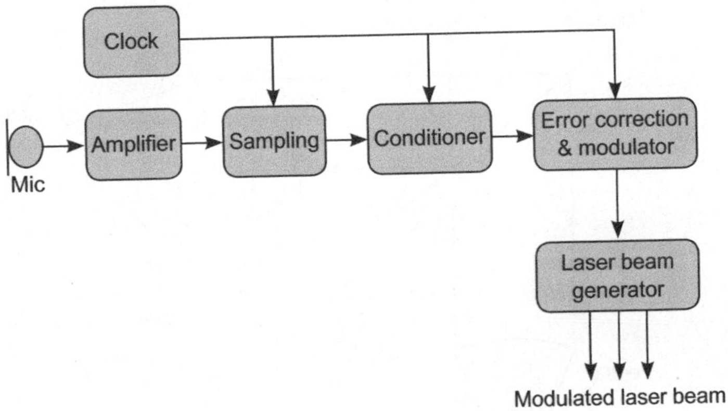


**Fig. 1.10** Block diagram of digital recording.

Moreover, digital audio is modulated by a powerful laser beam, and the resulting signal is used to record onto a Resist Mater Disc (RMD). The sampling frequency of the audio signal is 44.1 kHz. There are 705,600 bits per second, plus additional bits for error correction, signal control, and modulation. The laser beam's ON/OFF state is then modulated using the pulses. (Fig 1.10) displays the block diagram of the recording system. A laser beam is concentrated and then incident upon the master disc.

### 1.3.3.3 Optical Recording

Optical recording (Fig 1.11) involves making a laser beam incident on the CD via a half-silvered mirror. This mirror lets the input beam through but blocks the returning one. In this way, number one is represented by the reflected beam from the flat aluminium surface. For example, if a pit's reflection is very faint, it would be equal to 0. As a result, the laser beam that is reflected back to you is an exact reproduction of the original laser beam, with the audio signal represented as binary digits. The pit does not reflect any light, while the flat surface reflects it entirely. Therefore, when this ON-OFF reflected light strikes a photosensitive diode, binary digits are reproduced. A DAC is used to process the digital signal from the diode's output and return to its original analogues form.

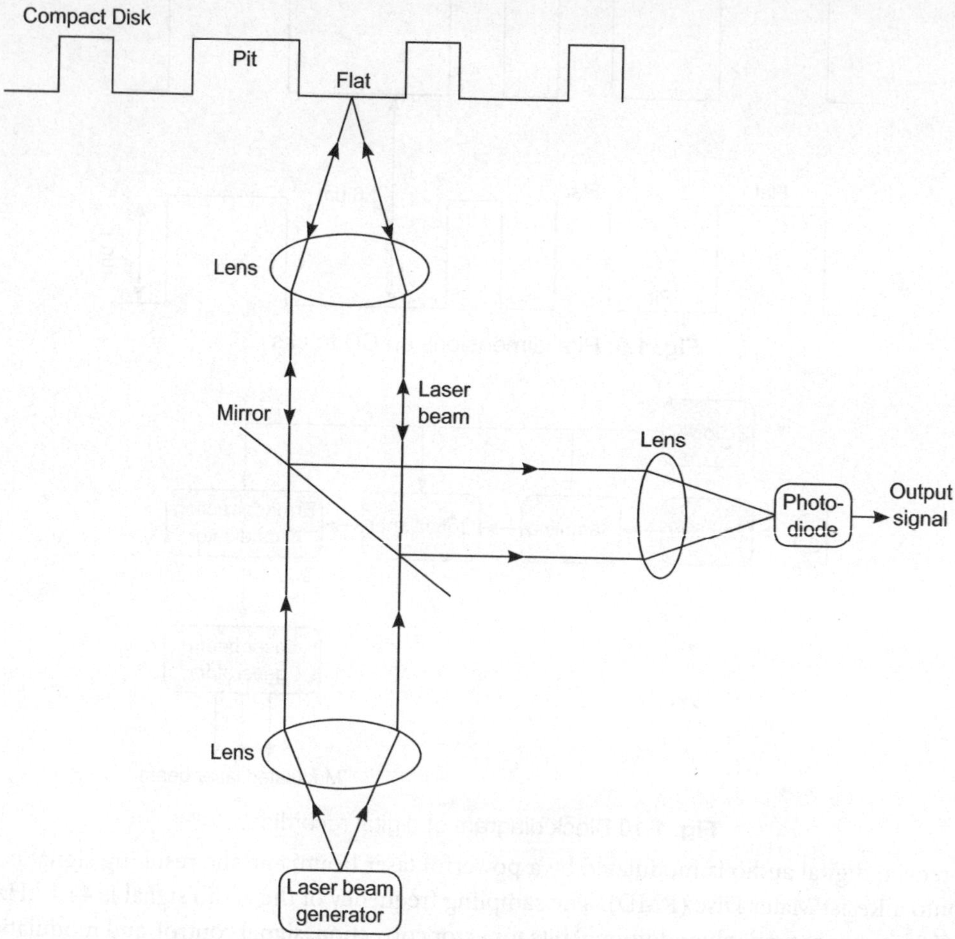


**Fig. 1.11:** Optical recording of sound

## UNIT SUMMARY

Standard audio level metres measure signal characteristics such as loudness and volume, pitch, frequency response, fidelity, sensitivity, and selectivity. The standard pitch frequency is 440 Hz, and it is used to describe the audibility of a sound. In audio, the frequency response is the range of frequencies that a component can faithfully reproduce. The term "fidelity" refers to the precision with which audio components and recordings can replicate the sound of an original recording.

Transducers known as microphones detect and record fluctuations in ambient sound levels via extremely small electrical currents. Microphones can be broken down into a wide variety of subsets, each of which serves a unique function. The PA system, broadcasting, military operations, and telephone industries rely heavily on carbon microphones because of their high output levels and durability. Moreover, they are the most effective solution because of the low manufacturing cost and simplistic style. The magnet, the diaphragm, and the coil are the main components of a moving coil microphone, which uses electromagnetic induction to detect and record sound. Wireless microphones, also known as radio microphones, use a low-power, frequency-modulated, very high-frequency (VHF) transmitter. These microphones can be placed directly on the amplifier without the need for an additional cable.

A loudspeaker is an electrical signal transducer that efficiently converts audio-frequency electrical signals into audible sound waves. They can be classified into three types: direct radiating, indirect radiating and multi-speaker systems.

Using magnetic recording, the fluctuations in sound pressure are captured and stored as simple magnets. It is predicated on the idea that iron oxide and other materials can be magnetised by exposure to a magnetic field and then maintain their magnetism indefinitely. The relatively close elimination of background noise in digitally recorded sounds motivated the development of the CD format. As an added step, a strong laser beam is used to modulate digital audio, and the resulting signal is recorded onto a Resist Mater Disc (RMD). In the optical recording, a laser beam is an incident on the CD from behind a half-silvered mirror which permits the incoming beam to pass while blocking the outgoing beam.

## EXERCISES

### Multiple Choice Questions

1. What type of wave motion is sound?
   (a) Transverse (b) Pulse type (c) Electromagnetic (d) Longitudinal
2. Pascals is a unit of __________
   (a) Frequency (b) Amplitude (c) Time period (d) Velocity
3. What is the typical frequency of a standard pitch?
   (a) 440 kHz (b) 4400 Hz (c) 440 Hz (d) 4.40 Hz
4. Which of the following is not a characteristic of ideal fidelity?
   (a) Low signal-to-noise ratio (b) High dynamic range
   (c) No spatial distortion (d) No frequency distortion
5. Which of the following microphones has the highest noise pick-up?
   (a) Ribbon microphone (b) Crystal microphone
   (c) Carbon microphone (d) Moving coil microphone

| ANSWERS | | | | | | | | | |
|---|---|---|---|---|---|---|---|---|---|
| 1. | (d) | 2. | (b) | 3. | (c) | 4. | (a) | 5. | (d) |

### Short and Long Answer Type Questions

1. Describe the following terminologies: amplitude, frequency, velocity, wavelength and phase.
2. Outline audio-level metering.
3. State the characteristics of ideal fidelity.
4. Differentiate between crystal microphones and ribbon microphones.
5. Classify the various types of speakers.

6. Describe the working principle of a microphone and state its characteristics.
7. Explain the moving coil microphone with the help of a suitable diagram.
8. Explain any four characteristics of a sound signal.
9. Describe the working principle of a speaker and state its characteristics.
10. Explain the digital and optical recording of sound.

## NUMERICAL PROBLEMS

1. Calculate the wavelength of sound at 900 Hz assuming its velocity is 344 m/s.
2. A car's sound has a pressure of $20 \times 10^{-3}$ Pascal and an intensity of 50 dB. Determine the sound pressure level needed to trigger a person's threshold for sound.
3. A sound pressure intensity of $20 \times 10^{-3}$ Pascal is 0 decibels. What will its intensity be at 0.2 Pascal Pressure?
4. Determine the velocity of sound at 273°C if its velocity at 0°C is 344m/s.

## PRACTICAL

1. To study the basic characteristics of a sound signal.
2. To study the working principle of a microphone. Understand the various characteristics of different types of microphones.
3. To study the working principle of a speaker.
4. To study the different methods of recording sound.

## KNOW MORE

- With his 1876 patent, Alexander Graham Bell created the first microphone. His mic was just a piece of wire that carried DC electricity (DC). A transmitter with a moving armature and a receiver was able to send and receive audio signals in both directions.
- Acoustician and inventor James Edward Maceo West was born in the United States. More than 250 U.S. and international patents cover his innovations in microphone construction and design, as well as his methods for making polymer foil electrets.
- The word "microphone" can be split into the components "micro" and "phone." Greek for "small" (micro) and "sound" (phone) combine to form the word "microphone." Because it picks up very faint sounds, the word microphone literally means "small sound."
- Johann Philipp Reis, a teacher in Friedrichsdorf, Germany, invented a basic electronic loudspeaker in 1861. The speaker's reproduction of the noise was rough, but it was an experiment.
- Oliver Lodge created the first unconventional moving-coil (or dynamic) loudspeaker in 1898. In 1915, in Napa, California, Danish engineers Peter L. Jensen and Edwin Pridham produced the first functional moving-coil loudspeakers.

## REFERENCES

1. Direct radiating loudspeaker, accessed on 10 November 2022, <https://in.rsdelivers.com/product/rs-pro/rs-pro-15w-grey-horn-speaker-ip66-250-hz-16-khz/5350316>
2. Indirect radiating loudspeaker, accessed on 10 November 2022, <https://www.indiamart.com/proddetail/dual-cone-loudspeaker-x300-15inch-8705002112.html>
3. Multi-Speaker System, accessed on 10 November 2022, <https://www.crutchfield.com/S-ZbSK2YTH4pY/learn/learningcenter/home/multiroom_power.html>
4. Gupta, R. G. (2010). Audio & Video Systems, 2e. Tata McGraw-Hill Education.
5. Crowhurst, N. H. (1969). Audio Systems Handbook. G/L Tab Books.
6. Adams, J. J. (1998). Complete Guide to Audio. Prompt.
7. Winer, E. (2017). The audio expert: everything you need to know about audio. Routledge.
8. Whitaker, J., & Benson, B. (2001). Standard handbook of audio engineering. McGraw-Hill Education.
9. Harris, K. D. (2007). U.S. Patent Application No. 29/224,646.
10. Self, D. (2012). Audio engineering explained. Routledge.

# 2 Audio Systems

## UNIT SPECIFICS

Through this unit, we have discussed the following aspects:

- Working of CD Player;
- Complete overview of Home Theatre Sound System;
- Understanding the Surround Sound System;
- Analysis of Digital Console Block;
- Diagram and Working Principle of FM Tuner;
- PA Address System working and applications;
- Speaker Impedance Matching;

The topics are discussed with various examples for generating further curiosity and creativity and improving problem-solving capacity.

Besides giving a large number of multiple-choice questions as well as questions of short and long answer types marked in two categories following lower and higher order of Bloom's taxonomy, assignments through several numerical problems, a list of references, and suggested readings are given in the unit so that one can go through them for practice.

After the related practical, based on the content, there is a "Know More" section. This section has been carefully designed so that the supplementary information provided in this part becomes beneficial for the users of the book. This section mainly highlights the initial activity, examples of some interesting facts, analogy, history of the development of the subject focusing on the salient observations and finding, timelines starting from the development of the concerned topics up to the recent time, applications of the subject matter for our day-to-day real life or/and industrial applications on a variety of aspects, case study related to environmental, sustainability, social and ethical issues whichever applicable, and finally inquisitiveness and curiosity topics of the unit.

## RATIONALE

The fundamentals of using a CD player, a sound system for a home theatre, a digital audio console, surround sound, and a public address (PA) system will be addressed in this module. A digital audio console's inner workings are broken down here, along with their various applications, to provide a deeper understanding of the device's operation. In addition, an explanation has been provided for each component of a public address system as well as the function that it serves. The process that must be followed in order to establish a public address system is variable and must be tailored to the particulars of each scenario, which have been outlined in the module.

## PRE-REQUISITE

Mathematics: Electronics Circuits (Class XII)

Physics: Signal and Wave (Class XII)

## UNIT OUTCOMES

Outcomes of this unit are: After learning this unit student will be able to

**U2-O1:** Explain CD players and Home Theatre including Surround Sound System

**U2-O2:** Describe the working of the Digital Console Block

**U2-O3:** Evaluate Public Address Systems as per design requirements

**U2-O4:** Design requirements of appropriate PA system for different applications

| Unit-2 Outcomes | EXPECTED MAPPING WITH COURSE OUTCOMES (1- Weak Correlation; 2- Medium Correlation; 3- Strong Correlation) | | | | | |
|---|---|---|---|---|---|---|
| | CO-1 | CO-2 | CO-3 | CO-4 | CO-5 | CO-6 |
| U2-O1 | 3 | 3 | – | – | 1 | 2 |
| U2-O2 | 2 | 2 | 1 | 2 | 1 | 1 |
| U2-O3 | 3 | 3 | 2 | 2 | 2 | 2 |
| U2-O4 | 3 | 2 | 1 | 1 | 2 | 2 |

## 2.1 CD PLAYERS

Numerous optical disc varieties exist, each with its own set of features. CDs, or Compact Discs, are mass-produced digital audio recordings aimed at consumers. Philips' strategy was to develop a new optical medium that, like the vinyl record, could be mass-produced by moulding or stamping without requiring the user to create a personal copy. Data is stored on it as bumps with flat tops embedded in a plastic sheet. These relief structures are typically too low-contrast to be read without the aid of phase contrast microscopy, which creates an optical interference effect to give the illusion of contrast.

The layer's reflectivity comes from a thin aluminium coating. Discs are read by focusing a pinpoint of light onto the data layer and measuring how its surface texture modifies the light's reflection back to the sensor. The optical disc track dimensions are extremely small. This concept is used in recordable CDs, where the relief structure is read like a regular CD. Evidently, a pattern of holes or blisters formation is irreversible once it has begun.

**Fig. 2.1:** CD Player[1]

## 2.2 HOME THEATRE SOUND SYSTEM

The Dolby-B system, designed specifically for use in home theatres, is a more standardized version of the original Dolby-A system. The Dolby-B system employs a single frequency band of encoding, from 500 Hz on up. Through the use of a filter, the low-level signals are boosted to the high-frequency range. The resulting signals feature a ten-decibel-increased amount of hiss and modulation noises. The Dolby output is the sum of the direct input signal and the signal that has been processed by the variable filter. To reproduce the signal, it must be inverted and sent through a continuously adjustable filter. The information that has been decrypted is then sent out of the adder after going through the filter.

**Fig. 2.2:** Typical Dolby-B system[2]

## 2.3 SURROUND SOUND

When using surround sound, additional channels are added from loudspeakers on the sides and behind the listener to give the impression that sound is coming from all directions on the horizontal plane (at ground level). In addition to providing a more immersive and cinematic listening experience, surround sound systems (depending on the system) typically provide higher-quality sound. Creating a surround sound effect in digital systems was made possible by artificially delaying the signal produced by the stereophonic system's two channels, thereby keeping the system's cost within reasonable bounds. The artificial delay enhanced the already present delay in echoes off the ceiling and walls. As a result, a surround sound experience was had at a reasonable price.

## 2.4 DIGITAL AUDIO CONSOLE

A basic audio analogue mixer takes in multiple input streams and merges them into one. Simply connecting all the input signals in parallel and feeding them into a single input won't work because they may interfere with one another. Separation of the signals is essential, allowing for independent regulation of each signal's volume. In addition to basic mixing, a mixing console's many features include phantom power for condenser/capacitor microphones, pan-control (which allows each signal to be placed in any position in the stereo field), filtering, equalisation, routing, and monitoring facilities, which allows any number of sound sources to be routed to the desired loudspeaker for listening without affecting the mixer's main output.

### 2.4.1 Working Principle

Professional audio engineers use a digital console to modify the characteristics of incoming audio signals. Digital signal processing allows for the flexible manipulation of acoustic properties such as mix, route, signal dynamics, and equalisation to meet a wide range of applications. Multiple audio channels, each with its own digital processing, are combined into a single "master" channel. Powerful signals in terms of control, processing power, and the desired audio quality can be generated with the help of a professional digital mixing console.

The audio mixing system's block diagram and power amplifier are depicted in (Fig 2.3). The FM console employs a TDA 7021T integrated circuit. It is a low-power FM radio receiver IC that operates on just 3 volts and needs almost no additional components. However, its low sound output necessitates an external amplifier which is IC LM386. The IC is protected against short circuits and operates over broad common-mode and differential voltage ranges.

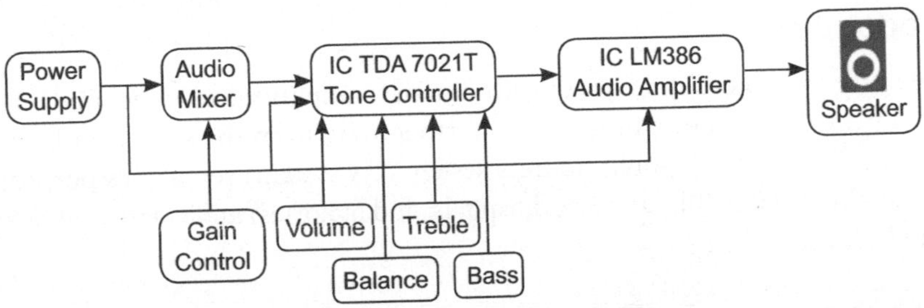


**Fig. 2.3:** Digital audio console block diagram.

### 2.4.2 FM Tuner

The mixer console allows transmission of audio from a computer, turntable, CD player, microphone, and hybrid phone without any interruptions or pauses, and it does so reliably and predictably every time. Turning a microphone on or off at a broadcast radio console will silence the broadcast and activate an "on-air" indicator. Some devices, such as CD players, phones, and computer channels, are set up to "trigger" a specific input when turned on. Audio in a digital audio console does not physically travel through the console's circuitry, as it does in an analogue audio console, which is why they are replacing them at many radio stations.

There are numerous radio transmitters in the 88-108 MHz frequency range, all of which can induce signal voltages in the aerial. The RF amplifier is able to pick out and amplify a certain station from among many, and its selection frequency can be changed as needed. This process is called tuning. The mixer receives both the specified frequency and the oscillator's output. A frequency-altering circuit is composed of a mixer and an oscillator. The Intermediate Frequency (IF) that comes out of the mixer is always at a consistent 10.7 MHz. An IF amplifier receives the input signal at its input. IF amplifiers are useful since their frequency and bandwidth are unaffected by the signal's frequency. This greatly simplifies the amplifier's design and operation. Moreover, the audio signal is reconstructed while the RF carrier is discarded by this circuit. A voltage amplifier is used to boost the volume of an audio signal by amplifying its voltage. The power amplifier boosts the signal sufficiently to power the speaker. The block diagram of a typical FM tuner is depicted in (Fig 2.4).

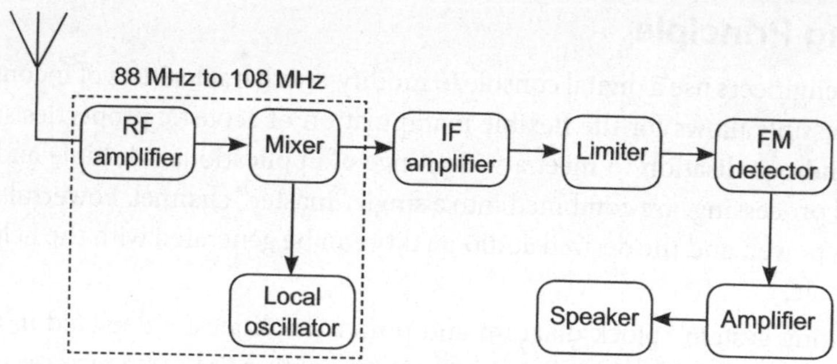


**Fig 2.4:** Block diagram of an FM tuner.

## 2.5 PUBLIC ADDRESS SYSTEM

### 2.5.1 Introduction

Because the intensity of a sound diminishes with distance, amplification is required whenever a large group is to be addressed; this allows those seated further away from the stage to still hear the speaker clearly and without discomfort. A public address system (PA system) serves this purpose. It is used to reach people in far-flung places (like airports, hospitals, and factories) and at large gatherings (like town hall meetings and concert halls).

### 2.5.2 Block Diagram of a Basic PA System

The microphone in a public address system converts acoustic signals into electrical ones. The loudspeaker is another transducer that transforms electrical audio signals into sound waves by amplifying and processing them. A typical PA system's block diagram is depicted in (Fig 2.5).

Components of a PA system are as under:

1. **Microphones**: Sounds are picked up and transformed into audio signals, which are merely variations in electrical current. Amplifiers typically feature multiple microphone inputs and an auxiliary input for a record player or tape player.
2. **Mixer**: A microphone's output is routed to a mixer. When multiple channels are being fed into a single power amplifier, the mixer stage is there to ensure that none of them interferes with one another. It can be a standalone plug-in or an integral part of the structure.
3. **Voltage amplifier**: The mixer's output is further amplified by this device.
4. **Processing circuits**: Tone controls (bass/treble controls) and a master gain control are features of these circuits.
5. **Driver amplifier**: Because of the voltage amplification it provides, the internal resistance of the subsequent stage is lowered when the signal is fed into it. In this way, it triggers the power amplifier to produce a greater output.
6. **Power amplifier**: It boosts the signal's strength. In order to eliminate even harmonics from the output and prevent the transformer core from becoming saturated, a push-pull circuit topology is used. A matching transformer connects the power amplifier's output to the loudspeaker so that the low impedance of the loudspeaker doesn't limit the amount of power that can be transferred to it.

7. **Loudspeaker**: It takes in electrical audio signals and transforms them into pressure variations, which we perceive as audible sounds.

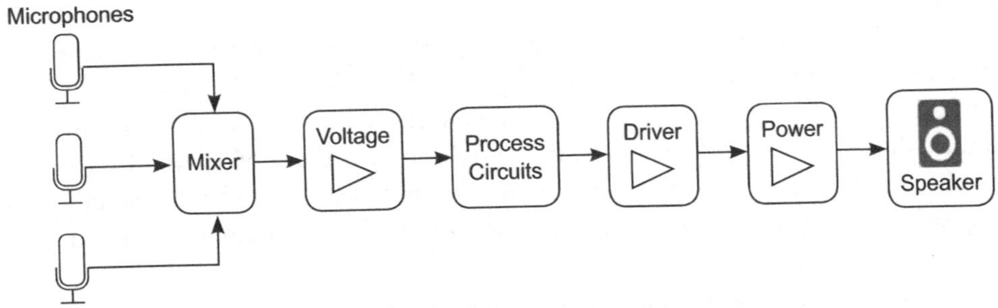


**Fig. 2.5:** Block diagram of a basic PA system.

## 2.6 PLANNING

### 2.6.1 Planning for Various Specification Requirements

1. **Acoustic Feedback:** In order to prevent an extremely loud howling noise caused by acoustic feedback, the microphone must be kept well away from the loudspeaker.
2. **Distribution of Sound Intensity**: The high notes are what allow us to understand what are being heard by us while the low ones provide the volume. Intelligibility decreases at greater distances because high notes attenuate more quickly than low notes when propagated further. Because of this, the sound pressure level needs to be split between multiple speakers so that it can reach even the farthest corners of the room.
3. **Reverberation**: Because sound waves in a reverberating medium overlap with one another, intelligibility is diminished. If the direct sound is getting lost in the echoes, the PA system should boost the volume there. Multiple low-power loudspeakers placed strategically around the auditorium are preferable to a single, large one because they reduce echo.
4. **Orientation of Loudspeakers**: Loudspeakers should be positioned so that the maximum output of the PA system is directed toward the audience and not the walls. In addition, they should be elevated one metre from the ground, with their axes roughly at the level of the seated audience members' ears. Reflections from faraway buildings can also cause audible echoes that need to be managed.
5. **Ambient Noise**: When there is a lot of background noise like in a busy market, the high frequencies (treble boost) need to be amplified in order to restore intelligibility. The high levels of amplification is required in a noisy environment which makes the sound seem artificial because bass is emphasised over treble. Therefore, in a noisy environment, the PA system should tone down the bass to preserve the quality of the sound. In addition, microphones with built-in noise reduction are useful.
6. **Dynamic Range**: Good public address systems have a level limiter built into the amplifier, which prevents the volume from rising above a certain threshold.

7. **Microphone selection:** The best microphones for a public address system are cardioid types, which pick up neither reflected sound nor sound from the loudspeakers themselves.
8. **The direction of arrival of sound**: The loudspeakers need to be positioned so that the sound appears to originate in that area. In general, the human ear can determine the source of a sound within a few seconds of hearing it. Accordingly, small speakers can be placed in such a way as to point listeners in the right direction, while larger speakers can be placed further apart to increase volume. Alternately, in digital systems, the signal can be artificially delayed by 10–20 ms before it reaches the loudspeakers, causing the loudspeakers to be heard after the sound directly from the speaker. Whatever the placement of the speaker, the desired volume and directionality will be achieved.
9. **Phase delay**: When multiple loudspeakers are playing at once, it is possible to hear both the sound coming from the closest loudspeaker and the other loudspeakers. When there is a delay of 45 milliseconds or more, the intelligibility of the delayed sound is compromised. The lag time is roughly 16 metres. Therefore, the maximum distance between speakers should not be more than 16 metres. When maintaining a safe distance, 10 metres is ideal.
10. **Grounding**: Earthing the chassis and shield of electronic devices and coaxial cables via plumbing is essential.
11. **Choice of Loudspeakers:** The selected loudspeaker must be able to cope with the amplifier's output power. Speaker columns with woofers, squawkers, and tweeters can be used to achieve high fidelity. Moreover, the efficiency of a public address system is improved by using horn-type speakers rather than cone-type speakers.
12. **Placement of Microphones**: Microphones should be positioned to pick up all of the program's sound without picking up any background noise.
13. **Speaker Impedance Matching**: Maximum power transfer from the amplifier to the loudspeaker is achieved by matching the total loudspeaker impedance to the amplifier's output impedance. Thus, the loudspeaker series-parallel combination should be optimized for maximum power transfer. For example, contemporary solid-state amplifiers, often known as "bridging" devices, take a voltage signal from an audio source and produce an amplified version of that signal. Output voltage and power are both dynamically regulated, and the output impedance is low.
14. **Power Amplifier:** The audience can hear us clearly from a distance thanks to the PA system's amplification. The amplifier's output power can range from a few watts for use in a classroom or at a small event to several hundred watts for use at a large public event or sporting competition.

### 2.6.2 Planning according to different situations

A badly installed PA system is very annoying and can ruin an otherwise enjoyable event. The setting where the public address system will be installed is a factor that must be considered during the planning phase. When making plans, these are the usual considerations:

#### 2.6.2.1 Public Meeting

(Fig 2.6) depicts a standard layout for an outdoor public meeting (120 m × 120 m). Most people in the crowd will be able to get their bearings thanks to the five pole-mountable loudspeakers (labelled A through E) in the centre of the stage. All the gathering areas will be covered by loudspeakers (letters

F through N). Furthermore, loudspeakers J and O are used to reach the far-off semi-circular side and corner sections. These would direct the loudness of the music into the walls. The areas to the left and right of the podium will be covered by the loudspeakers Q and R. S, T, V, X, Y, and Z loudspeakers cover the meeting park's outdoor audience. As depicted in the illustration, they could be slightly slanted. Microphones should be cardioid, while loudspeakers can be horn-type.

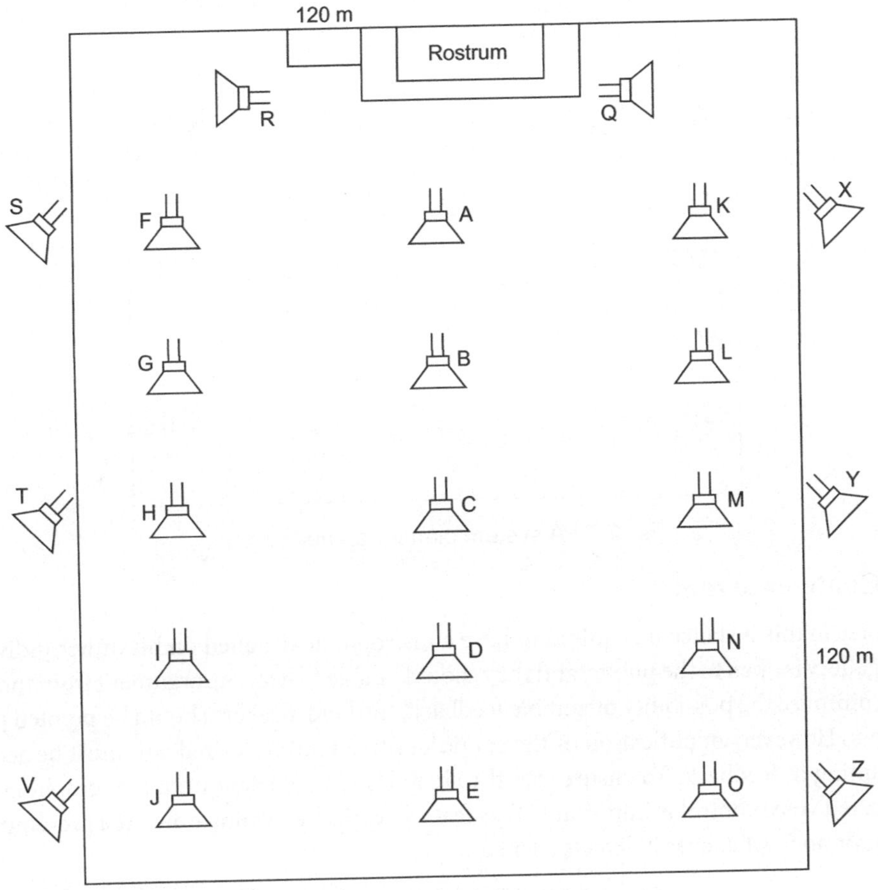


**Fig 2.6:** Installation plan of PA system for a public meeting.

### 2.6.2.2 Auditorium

A variety of events, including public gatherings, conferences, cultural programmes, etc., can be held in an auditorium. Therefore, the loudspeaker system needs to span a large frequency range (20Hz-20kHz) and have a high dynamic range (40dB-120dB). On either side of the stage, you should install a column of loudspeakers so that they point toward the audience. Moreover, having a dedicated, multifunctional mixer unit is preferable. It may accept a variety of microphones in addition to tapes and discs. Because of its potential for abuse in handling, moving-coil amplifiers are the best option. (Fig 2.7) depicts the layout of a typical auditorium. Additionally, the side speakers' axis should point towards the audience's ears rather than the floor or ceiling.

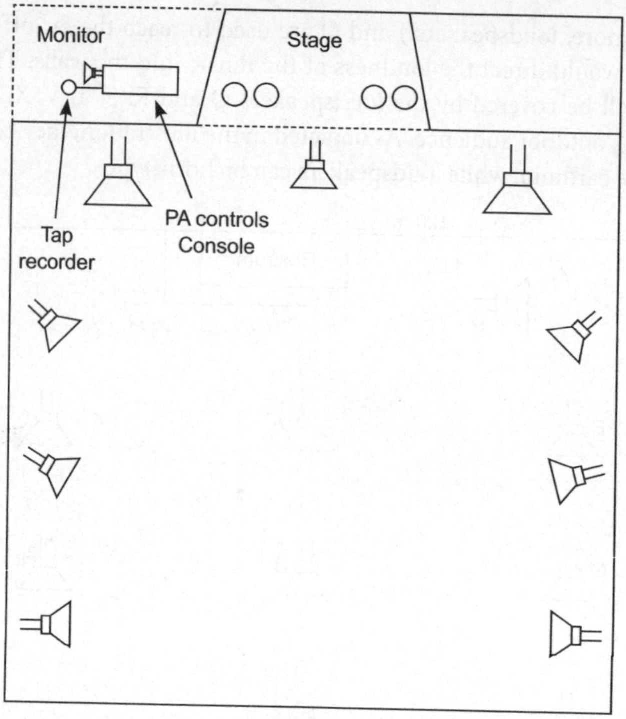


**Fig 2.7** PA system plan for an auditorium.

### 2.6.2.3 Conference room

Each presenter in this instance is required to have a microphone installed on his or her individual desk. The public gallery is open to the public, and they should be able to hear the argument from there. Further, in order to minimise the possibility of audible feedback, the loudspeakers should be pointed primarily at the spectators. However, amplification of the sound in some kind is desired and must be accomplished without resulting in feedback. To ensure that the channels remain silent until a microphone is actually used, automated voice control is important. This feature works by turning down the pre-amplifier of the relevant microphone whenever it detects silence.

### 2.6.2.4 Stadium

Stadiums typically have loud environments, therefore that should be taken into account. Columns placed in a row in front of each section of "spectators" can be tapped at 10W to provide shade for the standing spectators. If an amplifier were to fail, the sound system may be saved by wiring alternate columns to different amplifiers. Several amplifiers, each with a power output of 60 – 100 watts, should be present. Cardioid microphones are the preferred choice.

### 2.6.2.5 Sports Meet

Coverage of a sporting event is important for both the onlookers and the players on the field. In addition, the coverage is crucial so that the employees at the marquee enjoy the same level of excitement and amusement that the sporting event provides. Complete coverage of the sporting event can be achieved with a 60W amplifier.

## UNIT SUMMARY

In order to facilitate mass production by means of moulding or stamping, the optical disc was invented. Light is concentrated onto the disc's data layer, and the light's reflection off the disc's surface is measured to determine how the surface texture affects the light's return to the reader.

Developed with home theatres in mind, the Dolby-B system is ideal for those who want to watch movies at home. It uses encoding inside a single frequency band, from 500 Hz to higher. Dolby's output is the combined effect of the unprocessed input signal and the signal after it has been passed through a variable-passband filter. Surround sound is a method of reproducing an audio signal that gives the appearance that sounds are coming from all directions by using additional channels from loudspeakers placed on both sides and behind the listener. There is usually a noticeable improvement in audio quality when using a surround sound system. By intentionally delaying the signal produced by the two channels, surround sound effects were made possible in digital systems.

Control, processing power, and the desired audio quality can all be generated in large quantities by a professional digital mixing console. Mix, routing, signal dynamics, and equalisation are just a few examples of how sonic qualities can be manipulated by digital signal processing to fit a variety of uses.

To magnify sound, one should have the knowledge about as a public address system (PA system). With the use of amplifiers, public speakers can now address the big audiences in far-flung locations. Microphones, a mixer, a voltage amplifier, processing circuits, a driver amplifier, a power amplifier, and speakers are the main parts of a public address system. A public address system's setup procedure can change based on its requirements and the environment.

## EXERCISES

### Multiple Choice Questions

1. Which of the following is false for a Dolby-B system?
   (a) The signal is inverted before sending it to filter
   (b) It uses multiple frequency bands of encoding
   (c) The adder spits out the decoded data
   (d) Filter boosts the signal from low to high-frequency range
2. Which component is used to boost the low sound output of the digital audio console?
   (a) IC LM386 (b) Audio mixer (c) IC TDA 7021T (d) Gain control
3. Good PA systems have a level limiter which ________
   (a) Amplify ambient noise
   (b) Produce a delay of 45 milliseconds
   (c) Pick up ambient noise
   (d) Prevents the volume from rising above a certain threshold
4. The maximum distance between speakers should not be more than
   (a) 160 m (b) 0.0016 km (c) 0.016 km (d) 16 km

5. For an auditorium, the loudspeaker system should have a dynamic range of ________
   (a) 40 – 120 dB (b) 70 – 200 dB (c) 100 – 400 dB (d) 10 – 70 dB

| ANSWERS | | | | | | | | | |
|---|---|---|---|---|---|---|---|---|---|
| 1. | (b) | 2. | (a) | 3. | (d) | 4. | (c) | 5. | (a) |

## Short and Long Answer Type Questions

1. Write a short note on CD players.
2. Briefly describe the Dolby-B system.
3. Briefly describe a digital audio console.
4. State the components of a PA system.
5. Describe the planning of the PA system considering any 2 situations.
6. Explain the working of a digital audio console.
7. Describe the planning of the PA system for various specification requirements.
8. Explain the basic PA system with the help of a suitable block diagram.
9. Explain the installation plan of the PA system for a public meeting and an auditorium.

## PRACTICAL

1. To study the block diagram of digital audio console.
2. To study the working of a FM tuner. Understand the components of a FM tuner.
3. To study functions of the components of a PA system.
4. To study the planning of PA system based on various specification requirements and different situations.

## KNOW MORE

- A group of engineers from Sony and Philips collaborated in 1979 to develop the first digital audio disc. In 1980, the Red Book CD-DA standard was published following a year of testing and debate. Moreover, after their 1982 commercial debut, compact discs and players quickly gained widespread acceptance.
- In the 1920s, 16 mm projectors were used to create the first home theatres and in the 1930s, technological advances allowed for the creation of 8mm and sound 16 mm film.
- Once the cost of a Kodak 8 mm film projector dropped, showing home movies became increasingly common among middle- and upper-class American households in the 1950s.
- Surround sound was first used in a film in 1940 for the Disney animated short Fantasia. The operatic composition Flight of the Bumblebee by Nikolai Rimsky-Korsakov inspired Walt Disney to include a bumblebee in his musical Fantasia and make it seem as if it were flying in all corners of the theatre.

- When the REDD 17 stereo mixing system was built at London's Abbey Road Studios by EMI's Record Engineering Development Department in 1958, it marked a significant step forward in the recording industry.
- From 1911 through 1915, a series of experiments were conducted in which a microphone and loudspeaker were linked to a 12-volt battery, resulting in the first such instance of auditory feedback. Further studies led to the creation of 'Magnavox,' the first electric public address system.

## REFERENCES


1. CD Player, accessed 3 November 2022, <https://www.reevoo.com/p/sony-scdxe597-600975>
2. Dolby Atmos, accessed 3 November 2022, <https://www.bhphotovideo.com/explora/home-entertainment/tips-and-solutions/dolby-atmos-and-dtsx-what-are-they-and-how-do-i-get-them>
3. Gupta, R. G. (2010). Audio & Video Systems, 2e. Tata McGraw-Hill Education.
4. Harris, K. D. (2007). U.S. Patent Application No. 29/224,646.
5. Watkinson, J. (1997). Audio for Television. Routledge.
6. Crowhurst, N. H. (1969). Audio Systems Handbook. G/L Tab Books.
7. Winer, E. (2017). The audio expert: everything you need to know about audio. Routledge.
8. Whitaker, J., & Benson, B. (2001). Standard handbook of audio engineering. McGraw-Hill Education.

# 3 Television Systems

## UNIT SPECIFICS

Through this unit, we have discussed the following aspects:

- Monochrome TV standards;
- Aspect ratio of frame;
- Analysis and synthesis of picture details;
- Main components of the composite video signal;
- Colour TV standards;
- PAL system;
- Colour TV channels and transmitters.

The practical applications of the topics are discussed for generating further curiosity and creativity and improving problem-solving capacity. Besides giving a large number of multiple-choice questions as well as questions of short and long answer types marked in two categories following the lower and higher order of Bloom's taxonomy, assignments through a number of numerical problems, a list of references and suggested readings are given in the unit so that one can go through them for practice.

After the related practical, based on the content, there is a "Know More" section. This section has been carefully designed so that the supplementary information provided in thispart becomes beneficial for the users of the book. This section mainly highlights the initial activity, examples of some interesting facts, analogy, history of the development of the subject focusing the salient observations and finding, timelines starting from the development of the concerned topics up to the recent time, applications of the subject matter for our day-to-day real life or/and industrial applications on a variety of aspects, case study related to environmental, sustainability, social and ethical issues whichever applicable, and finally inquisitiveness and curiosity topics of the unit.

## RATIONALE

This fundamental unit of television systems demonstrates the foundations of monochrome and colour television operation. It explains the standards used in monochrome television, including the scanning process and video signal components. Scanning is an important process performed in a television system to obtain continuous frames and provide picture motion. Understanding how monochrome television works is essential because it serves as the foundation for colour television. Furthermore, the application of colour theory in colour television is described in detail. This describes the additive and subtractive mixing techniques used to create various colours. Furthermore, in order for television programmes to be received by the receiver, they must comply with specific transmission standards, which are discussed in detail. It also discusses the various TV cameras that have been developed. Finally, the transmission of TV channels is summarised.

## PRE-REQUISITE

Mathematics: Electronics Circuits (Class XII)

Physics: Signal and Wave (Class XII)

## UNIT OUTCOMES

Outcomes of this unit are: After learning this unit student will be able to

**U3-O1:** Describe the basic scanning process

**U3-O2:** Describe the components of a Video signal

**U3-O3:** Explain the colour theory for colour TV standards

**U3-O4:** Describe the different TV cameras

**U3-O5:** Explain the channels for TV transmission.

| Unit-3 Outcomes | EXPECTED MAPPING WITH COURSE OUTCOMES (1- Weak Correlation; 2- Medium Correlation; 3- Strong Correlation) | | | | | |
|---|---|---|---|---|---|---|
| | CO-1 | CO-2 | CO-3 | CO-4 | CO-5 | CO-6 |
| U3-O1 | – | – | 3 | 3 | 1 | 1 |
| U3-O2 | – | – | 3 | 3 | 1 | 2 |
| U3-O3 | – | – | 3 | 3 | 1 | 2 |
| U3-O4 | – | – | 3 | 3 | 2 | 2 |
| U3-O5 | – | – | 3 | 3 | 2 | 2 |

## 3.1 INTRODUCTION

A combination of the terms "tele" which means "far away," and "vision," which means "to see," gave rise to the term "television." People had been talking about seeing content at farther distances, but it couldn't be realized until J.L. Baird and others showed a conceptual demonstration between 1927 and 1930. After World War II, consumer markets had access to televisions that were at least tolerable, thanks to advancements like the scanning of picture tubes, beam deflection, visual amplification, and other related technologies. People of all ages and social groups have adopted television as a part of their family for a variety of reasons, including the capturing the information, education, and amusement.

Iconoscope is the first camera tube, which had the same working principle as cathode ray. The year 1930, marked the beginning of the use of electromagnetic scanning in imaging equipment such as cameras and television sets. The use of television as a medium for the dissemination of information and the provision of public enjoyment began to rapidly develop around the time when the Second World War came to an end. Because of this, there have been three distinct approaches to colour television developed, all of which are consistent with the three distinct criteria for black-and-white television.

Television has found its applications in human life in many ways including entertainment, education, news updates, advertisement, and public addressing (such as at airport/railway/bus terminals) etc. Closed Circuit Television (often abbreviated as CCTV) is a subgenre of television in which only a limited number of sites are permitted access to the camera feeds that are transmitted through cable connections. Users can use this feature to observe locations that they would not otherwise have access to for security or convenience reasons.

The quality of both transmission and reception saw significant improvements as a result of the development of transistors and integrated circuits. Because of vidicon and plumbicon, it was also possible to broadcast television shows in environments with low levels of illumination. Visuals on television are constructed using something called "pixels." The transition from analogue to digital technology made it possible to greatly increase the number of pixels in a picture, which is what drives up the resolution. The medium of television is always undergoing change. Televisions that use organic light-emitting diode technology (OLED) have the potential to produce astonishing effects, such as a high contrast ratio, thanks to electroluminescence. The introduction of the Smart TV platform made this achievement attainable. Applications for programming that are sent over the internet are typically bundled together with the product. OTT (Over The Top) platforms, such as Netflix and Amazon Prime Video, allow users to stream content to their devices. If in the next decade, every television can learn and make decisions on its own, virtual reality and future programming will have far more potential and power.

## 3.2 MONOCHROME TV STANDARDS

A monochromic image is composed of one colour and in the case of monochrome TV, images are black and white. With the C.C.I.R.(Consultative Committee for International Radio) system standards of 625 lines to be incorporated in India. Analogue television is the traditional television technology that uses analogue signals to transmit video and audio. The amplitude, phase, and frequency of an analogue signal represent the brightness, colours, and sound of an analogue television broadcast. An analogue television broadcast method is used to transmit visuals that are only in black and white. Amplitude modulation (AM) is performed on the visual signal along with the sound signal before transmission. Radiation of the modulated outputs, which are dispersed throughout the carrier frequencies, is accomplished with the help of an antenna. Receivers can be tuned to any station that is broadcasting because each station broadcasts on its own carrier frequency. A simplified block schematic of a TV transmitter and receiver systems are shown in (Fig. 3.1 (a) and 3.1 (b)), respectively.

### 3.2.1 Scanning Process

In today's motion pictures, 24 or more still photo images of the scene are taken every second and shown on the screen at the same rate. During the projection process, each image or frame is shown as a still sequential picture in rapid succession to create a cinematic effect. This provides the appearance that the scene being displayed is in motion at all times. Shutters are included inside the projector. As the film rotates in front of the light source, it can be projected onto the screen. During the transition to the next film, the film frame remains in place while the light from the screen is completely blocked out. Work is presently underway to position the frame in its final location. So, what you see on the screen is a rapid series of still film frames as a direct result of all of this. All light is blacked off during the transition between frames, allowing the eye to see clearly.

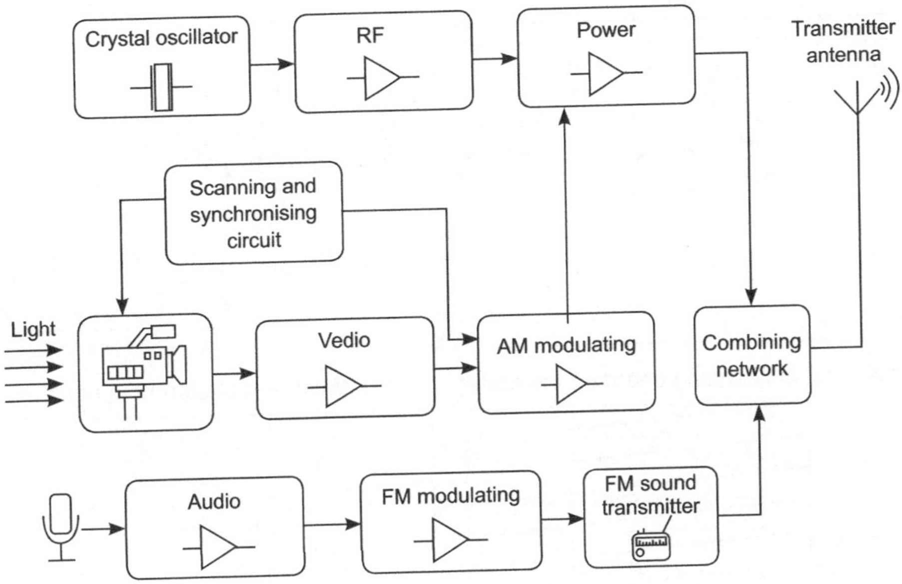


**Fig. 3.1 (a):** Basic monochrome television transmitter.

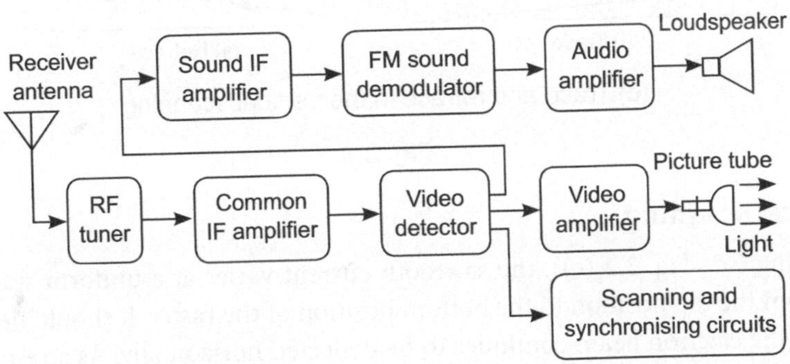


**Fig. 3.1 (b):** Basic monochrome television receiver.

An extremely comparable process can be found in the television set. As a result of this rapid scanning, the scene appears to be moving at a steady pace, even though it is being scanned in both horizontal and vertical directions, simultaneously. This allows individual elements of the picture to create an illusion that motion in the scene appears to be smooth and continuous. Typically, the repetition rate of a picture frame is kept at 25 per second. The scanning is broadly classified in two types: horizontal scanning and vertical scanning.

#### 3.2.1.1 Horizontal scanning

Horizontal scanning including trace as well as retrace of several horizontal lines is shown in (Fig. 3.2 (a)). In the horizontal deflection coils, (Fig 3.2 (b)) rise of the current ($i_H$) causes a continuous, uniform trace the period in which the beam is deflected from left to right across the screen. Once the peak of the rise is reached, the sawtooth current reverses its direction, resulting in a retrace (or flyback) period in which the current rapidly decreases and returns to its initial value. In summary, a horizontal scanning trace period is beam movement from left to the right on the raster, where the retrace movement is back to the left edge.

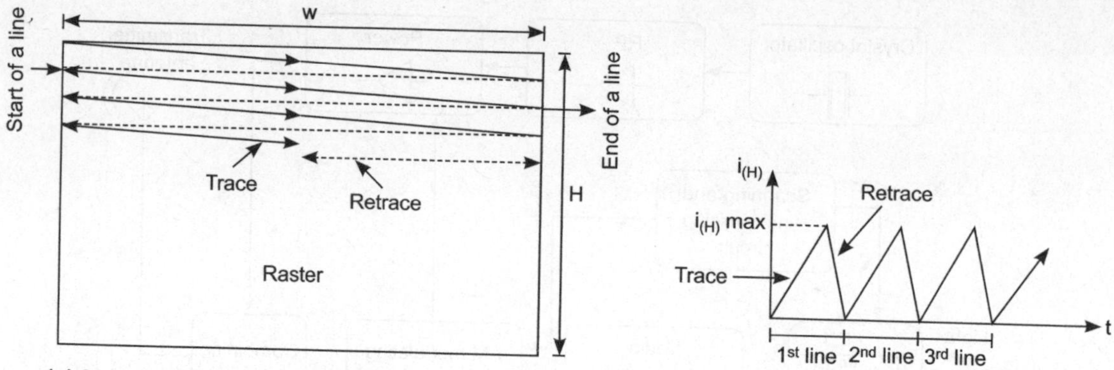


(a) Movement of scanning beam on the raster

(b) Waveform of current in horizontal scanning

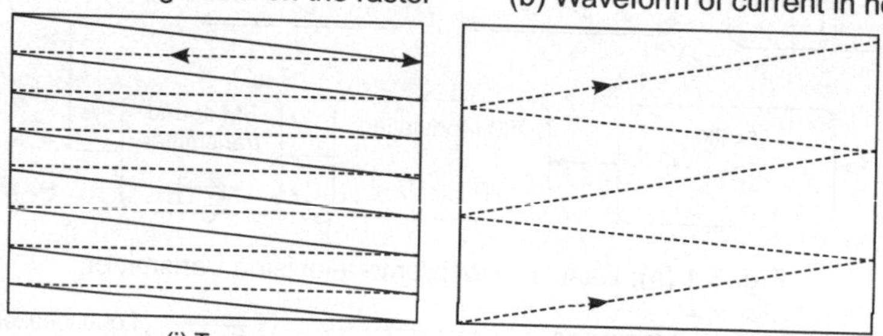

(i) Trace

(ii) Retrace

(c) Trace and retrace in the vertical scanning

**Fig. 3.2**

#### 3.2.1.2 Vertical scanning

In vertical scanning (see Fig. 3.2 (c)), the sawtooth current varies at a uniform speed and brings the electron beam from the top position to the bottom position of the raster. It should be noted that during vertical scanning the electron beam continues to be deflected horizontally. As a result, horizontal lines are produced one beneath the other while moving from top to bottom. When scanning vertically, the trace of the sawtooth wave deflects to the bottom of the raster. It is important to keep in mind that the scanning beams are blanked during horizontal and vertical retrace intervals, and no picture information is picked up or reproduced.

### 3.2.2 Aspect Ratio

The proportionate relationship that exists between the width and height of a picture is referred to as its aspect ratio. A shift in the picture's aspect ratio will result in a modification to the image's resolution, indicating that the resolution and aspect ratio are intertwined. The standard shape for the picture area in televisions is a rectangular one with a width-to-height (W2H) ratio of 4:3. It is referred to as the aspect ratio. For a raster screen, when the width (W) of the picture is kept more than its height (H), human eyes can view an image more comfortably. This is the reason that this particular choice of aspect ratio was made. Aspect ratios for modern television are often presented in the form of 16:9. Nevertheless, movies are made in accordance with certain cinematic principles. Therefore, the aspect ratio of motion pictures is 21:9.

### 3.2.3 Persistence of Vision and Flicker

The optical illusion of persistence of vision is created when an object has disappeared from view but the eye feels it can still see it because of how the eye works. As a consequence of this, there is a sensation that takes place whenever the retinal nerves are activated by light, and this sensation continues for approximately one-sixteenth of a second after the light has stopped being present. It is common practice to refer to this as the eye's capacity for storing information. When there are more images displayed to the eye in a given second or a higher rate of scanning per second, the human eye is better able to integrate the variations in brightness that are present in a scene. As a consequence of this, to the human sight, they appear to be a single image, with no noticeable differences between the numerous components that make up the image.

For creating the illusion of continuity, the motion picture rate of 24 frames per second (fps) rate and the scanning rate of 25 fps is adequate. However, it remains an issue of smooth blending of brightness frame to frame. This is because the screen is left blank in between each frame. This results in a distracting light flicker whenever there are alternating periods of bright and dark on the screen. The flicker is removed from motion pictures by displaying each frame of the film twice, although there are still 24 fps of moving pictures. One method that can be used to eliminate flicker in the video is called interlaced scanning.

### 3.2.4 Interlaced Scanning

With interlaced scanning, which makes use of 50 vertical scans per second, flicker-free television is now a reality.

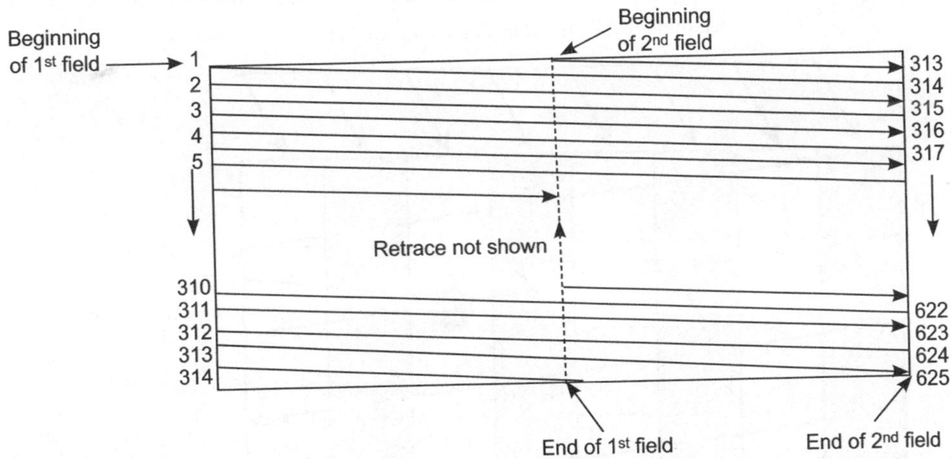


**Fig. 3.3:** Principle of interlaced scanning

In this scanning, the electron beam moves faster by scanning every alternate line rather than every line sequentially. As a result, it can read through more data in less time.

To finish its scan of lines that were missed during previous scans, the beam will reach the bottom of the picture frame. Total lines are divided into two independent groups known as "fields," and each field is scanned sequentially one after the other, as shown in (Fig. 3.3). The vertical retrace time is assumed to be zero.

### 3.2.5 Picture Resolution

Picture resolution is defined as the ability of the system to be able to produce the finest details of an object. Resolutions are defined separately for the picture's vertical and horizontal planes. Vertical resolution (Rv) refers to a scanning system's ability to present image details in the vertical direction. Vertical resolution in the 625-line system can then be expressed as

$$R_V = N \times k \tag{3.1}$$

where denotes the active number of lines and is the resolution factor.

Assuming $k = 0.69$,

$$R_V = 585 \times 0.69 = 400 \text{ lines} \tag{3.2}$$

The horizontal resolution (Rh) is determined by denoting the ability of a scanning system to present picture details in the horizontal direction. This can be evaluated by considering a vertical bar pattern as shown in Fig. 3.4. The aim is to have an equal vertical and horizontal resolution which is equal to

$$R_h = N \times \text{aspect ratio} = 585 \times 4/3 = 780 \tag{3.3}$$

The efficiency of a single line can change based on the number of black and white portions that are present in the image at any one point in time. As a result, when calculating the number of distinct picture components in a horizontal line, the same resolution factor can be applied to both vertical and horizontal resolution. This is due to the fact that equal vertical and horizontal resolution necessitates the same number of picture elements in each line. Given a horizontal line with the same vertical and horizontal resolution, the following equation can be used to determine the effective number of alternate black and white segments:

$$N_e = N \times \text{aspect ratio} \times k = 585 \times 4/3 \times 0.69 = 533 \tag{3.4}$$

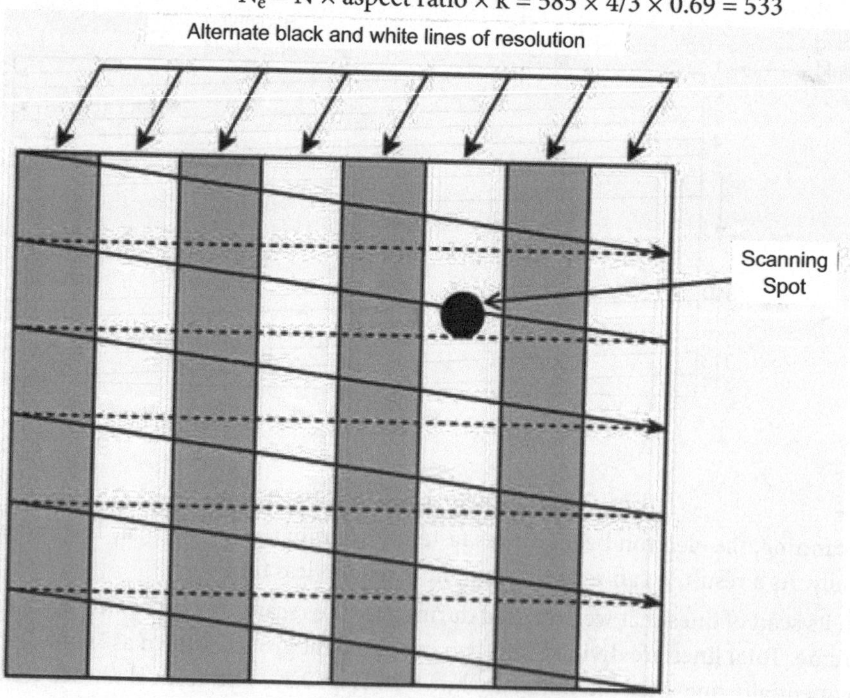


**Fig. 3.4:** Vertical bar pattern to determine amplifier requirements.

### 3.2.6 Composite Video (CVD) Signal

CVD is an analogue video signal format that transmits standard-definition video as a single channel. The CVD signal is made up of the following components:

- Video signal representing picture information details
- Blanking pulses to alleviate the effect of undesired retrace
- Sync pulses to achieve synchronization at transmitter and receiver sides

Sync pulses are sent one after other in time rather than simultaneously with the picture signal.

### 3.2.7 Horizontal and Vertical Sync Details

A horizontal synchronisation (sync) pulse is required at the end of each active line period, as well as a vertical sync pulse following the completion of each field scan. The amplitudes of the sync pulses are kept consistent to improve the effectiveness of the picture signal transmission. They are differentiated from one another at the receiver by a specific width, which results in a range of durations for each one.

#### 3.2.7.1 Horizontal Sync Details

H stands for the distance that separates each of the horizontal scanning lines. During the line blanking period, which lasts for 12 seconds and is a component of the entire line period of 64 seconds, a line synchronizing pulse is inserted. It makes use of an oscillator that can be synchronized thanks to pulses that correspond to the differentiated leading edges of sync pulses. There are three separate portions of the line blanking period that are referred to as "line sync" pulse and "back porch."

**Table. 3.1:** Horizontal Scanning time periods

| Event | Duration (ms) |
|---|---|
| Total lines (H) | 0.064 |
| Horizontal blanking | 0.012 ± 0.0003 |
| Horizontal sync pulse | 0.0047 ± 0.0002 |
| Front porch | 0.0015 ± 0.0003 |
| Back porch | 0.0058 ± 0.0003 |

The front porch comes before the sync pulse, and it allows the video circuit sufficient time to settle down to the blanking level from the currently present video level.

After the sync, the pulse has begun and the front porch of blanking has been completed, a horizontal retrace will be produced. Line sync pulses are split up at the receiver so that the receiver's line time base remains perfectly in sync with the line time base of the distant transmitter. In this amount of time, the backstroke (retrace) of the beam will almost certainly be finished, and it will arrive at the far left end of the raster. This is the shortest possible amount of time that it takes (4.7 μs).

When the blanking level is set to 5.8 μs, there is sufficient time for line flyback and reversal of current in the horizontal time-base circuit before scanning the next line can begin. When the blanking level is set to 5.6 μs, there is insufficient time for these processes.

#### 3.2.7.2 Vertical Sync Details

In both even and odd fields, the vertical sync added at the end is difficult to interpret. If the receiver is to be able to derive an acceptable field sync pulse for use in triggering the field sweep oscillator, the width of the field sync pulse must be maintained at a level much greater than the horizontal sync pulse. So the vertical sync duration is 2.5-3 times as long as the horizontal sync period as a result of this. In order to clearly distinguish horizontal and vertical pulses from each other, the width of vertical sync is kept sufficiently high.

Active and blanked lines have accessible sync pulses, however, the 2.5-line vertical sync period does not have any available leading edges. In both line periods, the horizontal sync pulses are accessible. Due to the fact that it operates at 15625 Hz, the horizontal sweep oscillator is more likely to go out of synchronism during each vertical sync interval.

### 3.2.8 Scanning Sequence

There are 625 lines per frame for scanning an image. This is divided into two parts: the first field (Table 3.2) and the second field (Table 3.3).

**Table 3.2:** First Field (odd field)

| Line numbers: 1st to the initial half of 313th line | | |
|---|---|---|
| 1, 2 and 3rd (initial half) lines | Vertical sync pulses | 2.5 lines |
| 3rd (later half), 4 and 5 | Post-vertical sync equalizing pulses | 2.5 lines |
| 6 to 17 and 18th (initial half) | Blanking retrace pulses | 12.5 lines |
| 18th (later half) to 310 | Picture details | 292.5 lines |
| 311, 312 and 313th (initial half) | Pre-vertical sync equalizing pulses for the 2nd field. | 2.5 lines |
| Total lines = 312.5 | | |

**Table 3.3:** Second Field (even field)

| Line numbers: 313th (later half) to 625 | | |
|---|---|---|
| 313th (later half), 314, 315 | Vertical sync pulses | 2.5 lines |
| 316, 317, 318th (initial half) | Post-vertical sync equalizing pulses. | 2.5 lines |

| | | |
|---|---|---|
| 318th (latter half) to 330 | Blanking retrace pulses | 12.5 lines |
| 331 to 623rd (initial half) | Picture details | 292.5 lines |
| 623 (latter half), 624 and 625 | Pre-vertical sync Equalizing pulses for the 1st field. | 2.5 lines |
| Total lines = 312.5 | | |

Interlaced scanning results in the V-sync pulse not being in the same location at the end of the two fields as it would be otherwise. In order to differentiate V-sync pulses from H-sync pulses that came before them, an integrating circuit is utilized. Post-equalizing pulses are what is utilized to counteract the effect that interlaced scanning has on the half-line. These pulses are introduced into the rear portion of the blanking pulse, shortly after the V-sync pulse and before the blanking pulse itself.

## 3.3 COLOUR TV STANDARDS

The first colour television sets were made commercially available to consumers in the year 1954. On the other hand, it did not become widely used until the middle of the 1960s. In a perfect TV system, the colours would be true to life, the brightness would be just right, and the definition would be sharp enough to be read by the naked eye. In spite of the fact that the image contains data on its brightness, there is a lack of colour information in certain parts of a monochrome picture. As a consequence of this, adding colour to the raster image that was previously white is essential. There are three characteristics of colour that can be used to characterize different aspects of visual information: luminosity, hue, and saturation.

### 3.3.1 Colour Theory

The creation of colour images requires the utilization of three bytes, which is equivalent to 24 bits of information. When a stimulus of adequate brightness is presented to the cones that are responsible for operational vision, the human eye is able to distinguish three primary colours: red, green, and blue. A total of eight bits are allotted to each colour in the pixel representation. The value '0' represents the absence of colour, also known as black, while the value '255' represents the colour white.

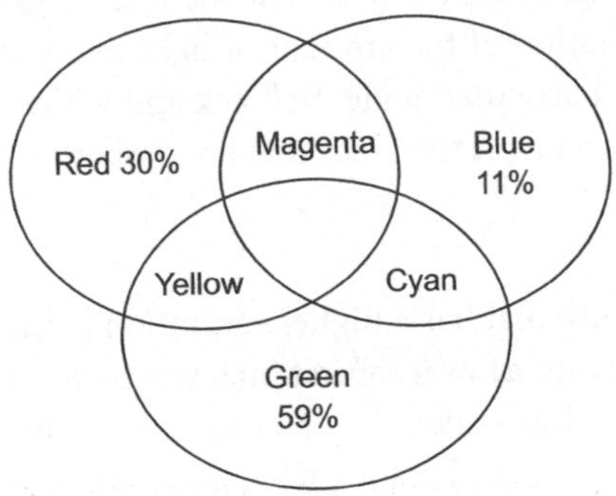


**Fig. 3.5:** (a) Additive colour mixing

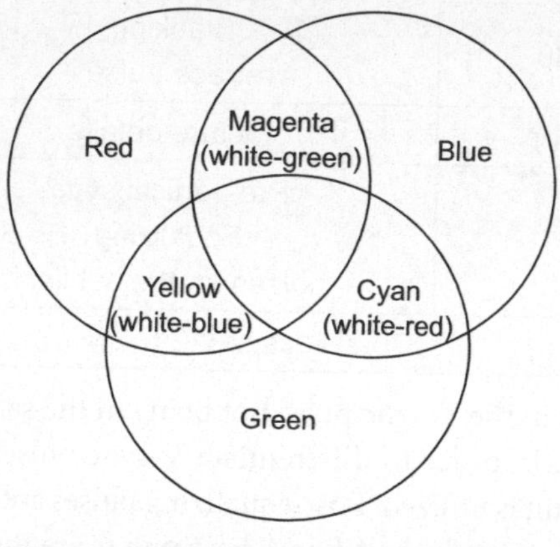


**Fig. 3.5:** (b) Subtractive colour mixing

The optic nerve system is responsible for determining the accurate colour of an object. Combining the several colour impressions can produce the desired effect, as demonstrated in the example given in Fig. 3.5. (a). For example, additive mixing is required for any colour television system to function. For example, if the green and blue cone groups are aroused at the same time with the same intensity ratio, a cyan colour may be seen if the cones are stimulated at the same time.

Subtractive mixing (Fig. 3.5 (b)) uses the reflective properties of pigments to limit the wavelengths of light absorbed by the pigments to those that correspond to the individual colours. When two or more different colours' pigments are mixed together, the result is a reflection of wavelengths shared by both colours.

### 3.3.2 HUE

It is the property of a colour that allows the colour of any given object to be distinguished from other colours. The cones in the retina are responsible for the perception of hue, which is determined by the varying wavelengths of the spectrum of radiation that are emitted. The term "tint" is another name for this substance.

### 3.3.3 Luminance

The amount of light that can be measured and is equivalent to brightness is referred to as luminosity. It provides a numerical representation of the amount of light that travels through or is emitted from a specific region when viewed at a particular angle. In black and white photographs, the sections that are better have a higher luminance level than the areas that are darker.

### 3.3.4 Saturation

When you apply a stain on a picture that has a higher saturation level, the image will appear more vivid and bright. When contrasted with this, a lower saturation level results in colours that have a more washed-out or faded appearance. Because white makes colours less strong, a higher saturation level indicates that colour has greater depth. A fully saturated colour does not contain any visible traces of white. Images that have a high saturation level are characterized by having vibrant colours that are saturated.

### 3.3.5 Chrominance

The term chrominance, often known as Chroma, refers to the combination of hue and saturation. However, it does not carry any information about the brightness of the object. The usage of chromaticity diagrams provides a depiction of all the spectral colours and their mixes that is clear to understand and is based on the primary colours that are contained within them.

### 3.3.6 Different Types of TV Cameras

A colour camera's job is to take an image and convert the intensity of each of the three main colours—red, green, and blue—into an equivalent amount of electrical voltage pixel by pixel. Instead of the brightness that is seen by the human eye, the voltage of the camera tubes is set up to match the actual brightness that is there in the colour. This is in contrast to how brightness is perceived by the human eye. As a direct consequence of this, the signals that were picked up are chrominance signals. There are two distinct varieties of camera tubes, namely the single-tube camera and the three-tube camera.

#### 3.3.6.1 Three-tube Camera

It is composed of three camera tubes that are identical to one another. These tubes can be vidicons, plumbicons, or any other form that is a modified version of these. They are the same kind of tubes that are used in monochrome television transmitters. The video voltage produced by colour cameras is used to indicate the absolute strength of each pixel, whereas the video voltage produced by monochrome cameras is used to represent how sensitive an eye is to each of its different colours. White balance is something that can be adjusted to make this possible.

#### 3.3.6.2 Single-tube Camera

The face plate of a single-tube camera features red, green, and blue stripes that are relatively close to one another. These stripes are where the colour filters are built in. The target, which likewise consists of stripes of the equivalent colour, is responsible for converting colour signals into the correct charge signal pattern. The scanning beam is responsible for the generation of visual signals by way of the corresponding load resistor. As a direct consequence of this, it is possible to observe video signals in the colours red, green, and blue.

## 3.4 TRANSMISSION STANDARDS

Television programmes can only be received by the receiver if it satisfies the requirements laid out by the transmission standards. As a result, it is necessary for the scanning processes utilized by both the receiver and the transmitter to be consistent with one another and coordinated. In a similar manner, for the receiver to be able to identify colour signals, it needs to generate the same subcarrier that was utilized by the transmitter. Different countries have developed their own unique sets of guidelines. As a consequence of this, receivers in a country are developed to be able to receive television transmissions that are in accordance with the standards of that country. Countries prefer minor changes in channel width, bandwidth, inter-carrier frequency, and the like because their scientists believe that minor changes will improve picture quality.

### 3.4.1 PAL System

A television encoding scheme known as Phase Alternate Line (PAL) is used to transmit analogue signals. 625 interlaced lines make up a PAL picture, which is shown in 25 fps. Colour systems that use an eye as an average mechanism for hue are called PAL colour systems. Above a certain threshold, the eye can see the effect of colour changes on alternating lines. As a result, this system must be modified. This can be improved by averaging the colours before displaying them to the eye.

This is because the PAL system has a higher number of scan lines. Stable colours are ensured through phase alternation error correction. The gamma and brightness signal bandwidths are also greater. Because it has more contrast than NTSC, another colour television standard.

The receivers, on the other hand, are expensive because they feature intricate electronic switching circuits. Compared to NTSC, the sound-to-noise ratio is lower with PAL. Phase differences cancel each other out, resulting in a reduction in colour saturation.

### 3.4.2 Channel Bandwidth

The frequency components of the video signal in the 625-line TV system range from 0 Hz to 5 MHz, and transmission of a Double Side Band (DSB) AM signal would necessitate a bandwidth of 10 MHz. This would not be enough to accommodate the transmission. Because it is hard to swiftly cut off the bandwidth of a signal at its sideband borders using filter characteristics that are appropriate, a band space that is even larger than it is already required. *When the bandwidth is raised, the speed at which* data can be carried may increase, and a greater number of devices may be connected all at once.

Fig. 3.6 depicts the entirety of a channel. Because the sound carrier is always at the very end of the completely radiated upper sideband, there is a 5.5 MHz gap between it and the image carrier. We can ensure that there will be minimal crosstalk between the two signals by placing them in this location. The FM sound signal that surrounds the sound carrier has a frequency range of about 75 kHz. To ensure proper inter-channel isolation, a guard band with a frequency of 0.25 MHz is permitted on the sound carrier side of the television channel. In comparison to the 11.25 MHz required by the corresponding double sideband signal, the entire channel bandwidth only takes up 7 MHz, resulting in a 4.25 MHz savings per channel.

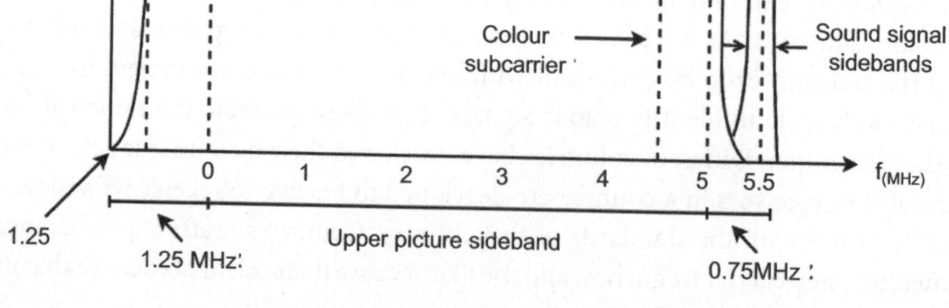


**Fig. 3.6:** Complete channel bandwidth

### 3.4.3 TV Signal propagation

Radio waves are electromagnetic waves that travel through space and are picked up by receiving antennas when they are released by antennas. Antennas are responsible for sending radio waves into space. There are three types of wave propagation based on the wavelengths of the waves involved: ground wave (or surface wave), sky wave, and space wave.

Electromagnetic waves with vertically polarised fields move toward the ground at very small angles. This is called "ground wave propagation," and the conducting surface of the earth guides them. This medium-wave transmitter is in charge of both long-wave telegraph and phone transmissions and medium-wave broadcasting.

Ground wave propagation becomes inefficient at frequencies above 1600 kHz due to signal attenuation caused by short-range transmission lag time. This occurs because of the lag time. As a result, sky waves are responsible for the overwhelming bulk of broadcasts in shortwave frequencies up to 30 MHz (11 meters). These waves travel in a direct line until they reach the ionosphere, where they are absorbed before continuing on their journey. As the frequency of the signal grows, the maximum angle at which it may be deflected back to Earth decreases until it is no longer possible to do so.

Radio waves with a frequency greater than 40 MHz cannot be carried by either surface or sky wave propagation (the start of the television broadcast band). As a result, the only option for transmitting signals in the VHF and UHF bands is to use radio waves that travel in a straight line from the transmitter to the receiver. The technical term for this phenomenon is "space wave propagation." For the signal to be propagated, the distance between the transmitter and the receiver must be in a direct line of sight.

### 3.4.4 Interference suffered by TV channels

There are numerous factors that interfere with carrier signals. These include thermal noise, man-made noise and various interfering signals originating from other similar sources of video or radio. Interferences can be categorized into three types:

#### 3.4.4.1 Co-channel Interference

In the event that two stations that use the same carrier frequency are situated in close proximity to one another, interference will occur. The phenomenon known as "venetian blind" interference emerges when two stations on the same co-channel are almost the same distance apart from one another. The visuals of the channels are displayed in black and white with a horizontal bar overlay that is also black and white. Interference levels cannot be ignored when the signal-to-interference ratio falls below 45 dB; at this point, these bars become more visible and move up and down the screen.

#### 3.4.4.2 Adjacent Channel Interference

Beats of frequencies generated by one channel can also generate interference in adjacent channels. The receiver antenna picks up a wide variety of additional signals in addition to the signal from the channel that is now selected. A significant portion of them is integrated in the RF tuner, so they are never even transmitted to the first IF stage. Interference signals from neighbouring channels can occasionally be so strong and so close to the channel of interest that they pass through the tuner section and appear at the first IF stage's input instead. The primary sources of interference are the various products produced by the various oscillator frequencies in conjunction with the sound carrier of the lower neighbouring channel and the picture carrier of the higher neighbouring channel.

On the display, a coarse dot pattern will appear any time the picture carrier of the desired channel is either quicker than the sound carrier of the lowest adjacent channel or faster than both of those carriers combined. If the lower neighbouring sound carrier is quite loud and the receiver does not muffle it suitably, beat patterns will become more noticeable.

**Table 3.4:** Television Channel Allocation (as per CCIR) in Bands I and III

| Band | Channel Number | Freq. Range (MHz) | Picture Carrier freq. (MHz) | Sound Carrier freq. (MHz) |
|---|---|---|---|---|
| I (41–68 MHz) | 1 | 41–47 | Not used | |
| | 2 | 47–54 | 48.25 | 53.75 |
| | 3 | 54–61 | 55.25 | 60.75 |
| | 4 | 61–68 | 62.25 | 67.75 |
| III (174–230 MHz) | 5 | 174–181 | 175.25 | 180.75 |
| | 6 | 181–188 | 182.25 | 187.75 |
| | 7 | 188–195 | 189.25 | 194.75 |
| | 8 | 195–202 | 196.25 | 201.75 |
| | 9 | 202–209 | 203.25 | 208.75 |
| | 10 | 209–216 | 210.25 | 215.75 |
| | 11 | 216–223 | 217.25 | 222.75 |
| Additional Channel | 12 | 223–230 | 224.25 | 229.75 |

### 3.4.4.3 Ghost Interference

Ghost interference is created by stray reflections of the signal from various structures such as building walls and roof surfaces. The reflected signal path is longer than the direct path, causing delayed arrival of the reflected signal. This delay is due to the greater direct signal regulating the synchronising circuitry, causing the image to be skewed. The greater direct signal causes this delay. In the industry, these displaced images are referred to as "trailing ghosts."

### 3.4.5 TV Broadcast Channels for Terrestrial Transmission

The most significant disadvantage of broadcast television is that it only has a limited audience. In order for signals to be received via terrestrial transmission, there must be a clear line of sight between the antenna that is transmitting the signal and the antenna that is receiving the signal. The greatest challenge presented by long-distance reception is posed by the spherical shape of the Earth, which causes the signal's line of sight to eventually become interrupted. As a result, satellite television is the superior choice because it is a wireless system that transmits programming straight to the customer's residence.

**Table 3.5:** UHF BAND IV (0.47–0.598 GHz) Channels 21–36

| Band (UHF) | Channel No. | Frequency Range (MHz) | Picture Carrier frequency (MHz) | Sound Carrier frequency (MHz) |
|---|---|---|---|---|
| IV Channel 21–36 | 21 | 0.47 - 0.478 | 0.47125 | 0.47675 |
| | 22 | 0.478 - 0.486 | 0.47925 | 0.48475 |
| | 23 | 0.486 - 0.494 | 0.48725 | 0.49275 |
| | 24 | 0.494 - 0.502 | 0.49525 | 0.50075 |
| | 25 | 0.502 - 0.510 | 0.50325 | 0.50875 |
| | 26 | 0.510 - 0.518 | 0.51125 | 0.51675 |
| | 27 | 0.518 - 0.526 | 0.51925 | 0.52475 |
| | 28 | 0.526 - 0.534 | 0.52725 | 0.53275 |
| | 29 | 0.534 - 0.542 | 0.53525 | 0.54075 |
| | 30 | 0.542 - 0.550 | 0.54325 | 0.54875 |
| | 31 | 0.550 - 0.558 | 0.55125 | 0.55675 |
| | 32 | 0.558 - 0.566 | 0.55925 | 0.56475 |
| | 33 | 0.566 - 0.574 | 0.56725 | 0.57275 |
| | 34 | 0.574 - 0.582 | 0.57525 | 0.58075 |
| | 35 | 0.582 - 0.590 | 0.58325 | 0.58875 |
| | 36 | 0.590 - 0.598 | 0.59125 | 0.59675 |

**Table 3.6:** UHF BAND V (606–870 MHz) Channels 37–69

| Channel Number | Freq. band (MHz) | Channel Number | Freq. band (MHz) |
|---|---|---|---|
| 37 | 606-614 | 54 | 742–750 |
| 38 | 614-622 | 55 | 750–758 |
| 39 | 622-630 | 56 | 758–766 |
| 40 | 630-638 | 57 | 766–774 |
| 41 | 638-646 | 58 | 774–782 |
| 42 | 646-654 | 59 | 782–790 |
| 43 | 654-662 | 60 | 790–798 |
| 44 | 662-670 | 61 | 798–806 |
| 45 | 670-678 | 62 | 806–814 |
| 46 | 678-686 | 63 | 814–822 |
| 47 | 686-694 | 64 | 822–830 |
| 48 | 694-702 | 65 | 830–838 |
| 49 | 702-710 | 66 | 838–846 |
| 50 | 710-718 | 67 | 846–854 |
| 51 | 718-726 | 68 | 854–862 |
| 52 | 726-734 | 69 | 862–870 |
| 53 | 734-742 | - | - |

### 3.4.6 Block Diagram of TV Transmitter

In the block diagram of the TV transmitter as shown in Fig. 3.7, each of the two types of transmitters—the amplitude-modulated transmitter and the frequency-modulated transmitter—plays a unique role. While an amplitude-modulated transmitter is used to modify the video, a frequency-modulated transmitter is used to alter the audio. Modulation of the video takes place.

A radio frequency (RF) carrier frequency is produced by the master oscillator in either scenario. When it detects the correct carrier, it produces a multiple of that carrier and drives the harmonic generators (frequency multipliers). A harmonic generator's output-tuned circuit is tuned to one of the input signal's harmonics. Harmonic generators are tuned amplifiers of type C. In a video transmitter, the carrier signal is fed into an amplitude modulator, whereas in an audio transmitter, the carrier signal is fed into a frequency modulator. The appropriate level of amplitude is transmitted with this signal as it enters the modulator. Because of the low-level modulation, linear amplifiers are used to increase the level of the modulating signal so that it may be sent successfully at the required level. After being combined on separate carriers, the video and audio are combined into a single signal before being transferred to the transmitting antenna.

## UNIT SUMMARY

Monochrome television displays images that are only shades of grey between black and white. To obtain continuous frames and provide picture motion, the scene is scanned in both the horizontal and vertical directions at a rapid rate. A typical television picture frame has a 4:3 aspect ratio. Furthermore, the motion picture rate of 24 fps and the scanning rate of 25 fps are used to avoid visual persistence. Flicker is caused by a blank screen between each frame. This can be avoided by employing interlaced scanning.

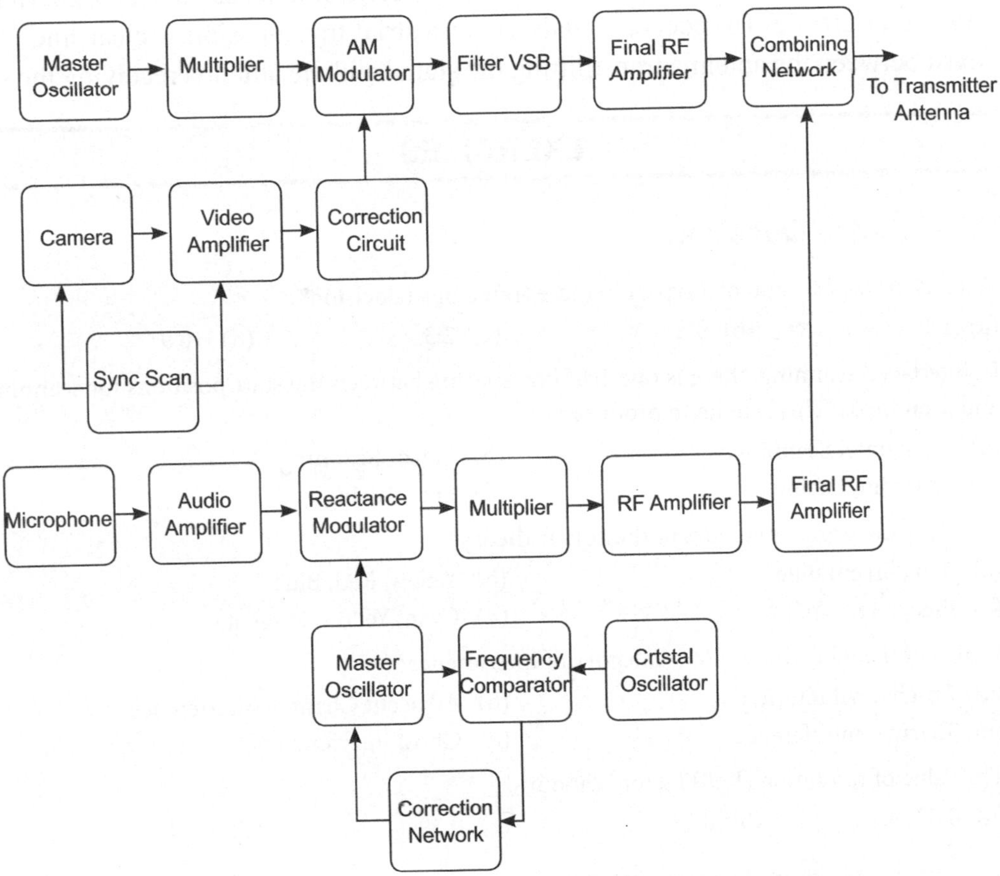


**Fig. 3.7:** Block Diagram of TV Transmitter

Despite the fact that the image contains brightness information, there is a lack of colour information in certain parts of a monochrome image. As a result, it is critical to add colour to the previously white raster image. As a result, there are three colour characteristics that can be used to characterise various aspects of visual information: luminosity, hue, and saturation. To create a variety of colours, additive and subtractive mixing techniques are used. The three primary colours used in these techniques are red, green, and blue. There are two distinct varieties of camera tubes, namely the single tube camera and the three-tube camera.

Television programmes can only be received by the receiver if they meet the transmission standards. Analogue signals are transmitted using the Phase Alternate Line (PAL) television encoding scheme. There are three types of TV signal wave propagation based on the wavelengths of the waves involved: ground wave (or surface wave), sky wave, and space wave. Furthermore, there are numerous factors that interfere with carrier signals, which can be classified into three types: co-channel interference, adjacent channel interference, and ghost interference. To receive signals via terrestrial transmission, a clear line of sight must exist between the antenna transmitting the signal and the antenna receiving the signal.

## EXERCISES

### Multiple Choice Questions

1. What is the aspect ratio of a typical picture frame in a television?
   (a) 1:1 (b) 4:3 (c) 2:3 (d) 16:9
2. In interlaced scanning, there is one-half line spacing between the start positions for scanning odd and even fields. This is done to produce
   (a) Horizontal Scanning (b) Linear Scanning
   (c) Exact Interlacing (d) Line Pairing
3. What are the primary colours of the colour theory?
   (a) Red, Green, Blue (b) Yellow, Red, Blue
   (c) Blue, Green Yellow (d) Cyan, Yellow, Magenta
4. Which of the following is not a category of interference?
   (a) Co-channel interference (b) Adjacent Channel interference
   (c) Thermal interference (d) Ghost interference
5. The value of resolution (Kell) factor is about _____
   (a) 0.67 (b) 0.71 (c) 0.59 (d) 0.69

| ANSWERS | | | | | | | | | |
|---|---|---|---|---|---|---|---|---|---|
| 1. | (b) | 2. | (c) | 3. | (a) | 4. | (c) | 5. | (d) |

## Short and Long Answer Type Questions

1. Explain in brief, the role of scanning in television systems.
2. Explain the following terms in the context of Television: flicker, Aspect Ratio, Blank pulse, Front and Back porch.
3. Explain why 625 lines were chosen rather than 623 or 627, and why the frame reception rate was set at 25 rather than 24 as in motion pictures.
4. Determine the width and height of the TV screen for a 60 cm TV with a 4:3 aspect ratio.
5. State the important cause of ghost images on TV.
6. Draw a block diagram and explain the monochrome TV transmitter and receivers.
7. Explain with necessary details Composite Video (CVD) Signal including Horizontal and Vertical Sync.
8. Explain the interlaced scanning process with the necessary details.
9. Mention important factors which are considered for designing the transmission and reception systems to get a good quality picture. Draw relevant current waveforms.
10. Describe the separation and processing of sync signals from incoming video signals in the TV receiver.
11. Explain with necessary details the PAL system and its system bandwidth.
12. Draw the block diagram of the Colour TV transmitter. Explain the same in brief.

## PRACTICAL

1. To study block diagrams of monochrome TV transmission and reception.
2. To study the function of the RF tuner section. Observe the necessary output waveforms & measure necessary test point voltages.
3. To study the video IF section & sound IF section and observe the CCVS signal. Observe AGC & AFT output.
4. To study the horizontal oscillator section and vertical oscillator section with the operation of Sync Separator. Observe the frequency of oscillation on various test points.

## KNOW MORE

- Although the requirements for monochrome transmission had been established before WWII, progress in civilian electronics was halted for the duration of the conflict.
- When television first debuted in 1936, it aired in black and white for the next 30 years, until 1967. Also, the first mechanical television broadcasts (1929–1935) only had a few thousand viewers, and the images they saw were black and orange because of the neon gas used in the lights.
- Scottish innovator John Logie Baird created the first successful colour TV systems in 1928. Based on Baird's 1928 mechanical TV concepts, CBS engineers developed a colour television system in 1940. The original colour televisions had awful picture quality and were not backwards-compatible with monochrome sets.

- In 1942, a young Mexican inventor called Guillermo González Camarena was granted a patent for a colour television. His invention, a "enhanced chromoscopic adapter" for Trichromatic-based colour TV transmissions, was granted a patent. He pioneered the use of colour on television and became widely recognised for this achievement.

## REFERENCES


1. Gulati, R. R. (2005). Monochrome and colour television. New Age International.
2. Gupta, R. G. Audio & Video Systems, 2e. Tata McGraw-Hill Education, 2010.
3. Yang, K., Li, Z., Shang, L., & Yi, W. (2011). Design and analysis of a novel surface acoustic wave micro position sensor. Journal of Electromagnetic Analysis and Applications, 2011.
4. Bhikshapathy, M. B., Pandharipande, V. M., & Krishna, P. G. Analysis & Design of Portable Video Transceiver.
5. Alkhatib, M. (2016). A New Optimized Double Stacked Turnstile Antenna Design.
6. Patel, C., & Kulkarni, G. R. (2010, December). Simulation and performance analysis of NTSC television system and standards. In 2010 IEEE International Conference on Computational Intelligence and Computing Research (pp. 1-5). IEEE.
7. Trundle, E. (2001). Newnes guide to television and video technology. Newnes.
8. Whitaker, J. C., & Benson, B. (2003). Standard handbook of video and television engineering. McGraw-Hill Education.
9. Tozer, E. P. J. (2012). Broadcast engineer's reference book. Routledge.
10. Fink, D. G. (1957). Television engineering handbook. McGraw-Hill.
11. Jack, K., & Tsatsoulin, V. (2002). Dictionary of video and television technology. Gulf Professional Publishing.
12. Goldmark, P. C. (1949). Brightness and contrast in television. Electrical Engineering, 68(3), 237-242.
13. Sharma, K. K. (2009). Fundamentals Of Television And Radar Engg. Seagull Books Pvt Ltd.
14. Dhake, A. M. (1980). Television Engineering (CCIR System-B Standards). McGraw-Hill Companies.
15. Roberts, R. S. (Ed.). (1985). Television Engineering: Broadcast, Cable, and Satellite. Pentech Press.
16. Srivastava, H. O., & Jain, R. C. (1999). Broadcast Technology—Past, Present and Future: A Review. IETE Technical Review, 16(3-4), 317-334.


□□

# 4 Television Receivers and Video Systems

## UNIT SPECIFICS

Through this unit we have discussed the following aspects:

- Basic components of PAL-D TV;
- Various parts of TV;
- Recent types of TV and their functionality;
- High-Definition Television (HDTV) standards;
- Direct to Home (DTH) working;
- Types of videos;
- Digital Videos.

The practical applications of the topics are discussed for generating further curiosity and creativity and improving problem-solving capacity.

Besides giving a large number of multiple-choice questions as well as questions of short and long answer types marked in two categories following the lower and higher order of Bloom's taxonomy, assignments through a number of numerical problems, a list of references and suggested readings are given in the unit so that one can go through them for practice.

After the related practical, based on the content, there is a "Know More" section. This section has been carefully designed so that the supplementary information provided in this part becomes beneficial for the users of the book. This section mainly highlights the initial activity, examples of some interesting facts, analogy, history of the development of the subject focusing the salient observations and finding, timelines starting from the development of the concerned topics up to the recent time, applications of the subject matter for our day-to-day real life or/and industrial applications on variety of aspects, case study related to environmental, sustainability, social and ethical issues whichever applicable, and finally inquisitiveness and curiosity topics of the unit.

## RATIONALE

This unit of television systems illuminates the concepts of television receivers and video systems. The human eye can not perceive the effect of changes in the "simple PAL" colour system hence for improvements it requires modifications, and this modified system is known as the "PAL-D" ( PAL -Delay Line) method which uses a delay line to compute averaging before presenting the colour to the eye. This unit describes how the PAL-D colour TV receiver functions in depth.

Furthermore, televisions have evolved significantly over the years. Digital, LCD, LED, Plasma, HDTV, 3-D TV, and Projection TV has been briefly described, as well as their benefits and drawbacks stated. Additionally, the unit demonstrates the operation of the outdoor and indoor units of direct-to-home (DTH) technology.

The television receiver includes a video amplifier, sound section, sync separator, and processor that have been thoroughly discussed. In addition, various types of video cables and interfaces have been discussed. Finally, the unit incorporates fundamentals of CD and DVD players.

## PRE-REQUISITES

Mathematics: Electronics Circuits (Class XII)

Physics: Signal and Wave (Class XII)

## UNIT OUTCOMES

List of outcomes of this unit is as follows: After learning this unit student will be able to

**U4-O1:** Describe PAL-D Colour TV Receiver Block Diagram along with its components

**U4-O2:** Describe DTH Concept Receive Block Diagram

**U4-O3:** Explain recent types of TV and their functionality

**U4-O4:** Explain interfaces and players of TV

| Unit-4 Outcomes | EXPECTED MAPPING WITH COURSE OUTCOMES (1- Weak Correlation; 2- Medium correlation; 3- Strong Correlation) | | | | | |
|---|---|---|---|---|---|---|
| | CO-1 | CO-2 | CO-3 | CO-4 | CO-5 | CO-6 |
| U4-O1 | 2 | 2 | 3 | 3 | 3 | 2 |
| U4-O2 | 1 | 1 | 3 | 3 | 2 | 1 |
| U4-O3 | 3 | 1 | 3 | 3 | 3 | 2 |
| U4-O4 | 2 | 2 | 3 | 3 | 3 | 2 |

## 4.1 INTRODUCTION

A television receiver combines AM and FM systems - AM for picture and FM for sound reception. Furthermore, the receiver circuitry includes scanning and synchronising blocks for clear image reproduction at the picture tube screen. The antenna section of the TV intercepts incoming signals and forwards picture plus sound carrier signals to the RF tuner. A TV receiver is a heterodyne type consisting of two or three phases of intermediate frequency (IF) amplifiers. The video signal is recovered by demodulating the output of the final IF stage. Picture information-carrying signal is driven to the picture tube, where picture elements are retrieved from electrical signals.

## 4.2 PAL-D COLOUR TV RECEIVER BLOCK DIAGRAM

Modern TV receiver circuit diagrams rarely allow for detailed signal tracing because of the widespread use of integrated circuits, which perform multiple functions at once. Although this demands a significant amount of circuit design standardisation, a wide variety of integrated circuits are available for colour TV receivers due to the many permutations of the functional units. Among the various PAL-D Colour TV Receivers that have been developed, The schematic representation of a commonly used setup is shown in Fig. 4.1. The overall signal flow pattern of a PAL-D colour TV receiver is similar to that of an NTSC receiver. The block diagram is divided into the following sections:

### 4.2.1 Tuner

A TV receiver tuner employs an additional circuit known as automatic frequency tuning (AFT) to retain the correct local oscillator frequency. In order to obtain perfect colour reproduction, a colour burst is required and in turn, it depends on AFT. Moreover, a picture IF is achieved at 38.9 MHz. The discriminator part of AFT processes IF signals to generate a dc control voltage proportional to any deviations observed in frequency. Error voltage, if any is passed the local oscillator to keep its frequency stable.

### 4.2.2 Sound Strip

To generate audio output, FM sound IF signal is processed. Volume, as well as tone controls of the sound signal, are first given to the loudspeaker via an audio amplifier.

### 4.2.3 Automatic gain control (AGC), Sync-separator & Deflection Circuits

For better reproduction of signals to the TV, the AGC circuit controls the gain of the receiver circuits in accordance with the strength of the signal received by the antenna. The sync separator is an important section to separate the sync pulses (which include horizontal, vertical, and equalising pulses) from the incoming video signal.

### 4.2.4 Luminance Channel

In the luminance channel, the first section is the video amplifier to amplify the signal. The video amplifier is followed by a delay line that is required to synchronise with the chroma signal that suffers delay due to the limited bandpass. Subsequently, there exists a notch filter that attenuates subcarriers by about 10db and reduces the appearance of any dot structure on the screen. At last, an inverted composite video signal available at the output is fed to the picture tube's three cathodes with proper brightness and contrast adjustments.

### 4.2.5 Colour Signal Processing

The video signal present at the video detector is amplified by a video preamplifier and to different parts. Nowadays integrated circuits (ICs) are used to process the colour signal.

### 4.2.6 Subcarrier Generation and Control

This section locally generates a frequency subcarrier to compensate for the loss of the subcarrier in the modulator section at the transmitter side. It is highly desired that the generated subcarrier should have frequency and phase references be the same as the original values at the transmitter. The auto phase control (APC) circuit controls the task by comparing the incoming bursts to the locally generated reference subcarrier. This ensures that the crystal oscillator in the receiver operates at the correct frequency and phase.

## 4.3 DIGITAL TV

Digital television (DTV) employs digital encoding to transmit television signals, in contrast to traditional analogue television technology, which uses analogue signals. It was the first major change to television technology since the emergence of colour in the 1950s, so its creation was greeted with excitement and anticipation. The widescreen aspect ratio (typically 16:9) is used, which is different from the more conventional 4:3 employed in analogue television. Compared to analogue TV, it is more efficient in its use of the radio spectrum, as it can transmit up to seven channels using the same amount of bandwidth as a single analogue channel.

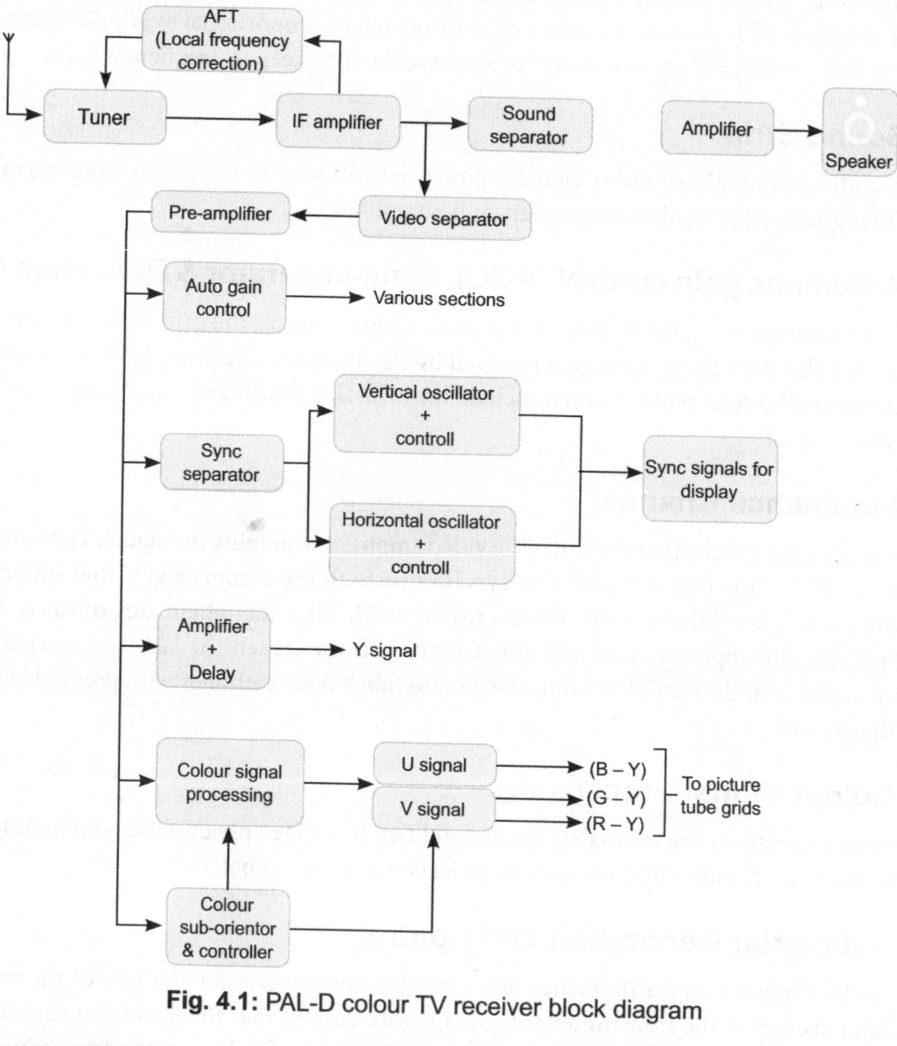


**Fig. 4.1:** PAL-D colour TV receiver block diagram

### 4.3.1 Advantages

1. Offers more channels as compared to terrestrial analogue
2. Better image resolution
3. Superior sound quality

4. If the television has a network and a broadband connection, numerous interactive services are available
5. Compatible with computers and the internet
6. Provide consistent reception over varying distances

### 4.3.2 Disadvantages:

1. Requires good-technical skills to set up a connection
2. The absence of properly designed guidelines may damage the quality.

## 4.4 LCD TV

In the early 21st century, sales of liquid crystal display (LCD) televisions skyrocketed, eventually surpassing those of CRT televisions worldwide in 2007. Moreover, LCDs were originally developed for use in portable televisions. LCD (liquid crystal display) TV utilises light-based modulating properties of liquid crystal in addition to polarizers. For images to be displayed on a liquid crystal display, a backlight or reflector is required. Consequently, they consume much less power than plasma displays.

**Fig. 4.2:** LCD TV [1]

Moreover, it consists of a grid of individual picture elements (pixels) that can be moved around and rearranged to create a new visual. Both passive and active-matrix display grids can be used in an LCD screen. Both the active-matrix LCD (also called a thin film transistor (TFT) display) and the passive matrix LCD (also called a dot matrix LCD) use a grid of conductors to organise their displays, with individual pixels placed at the grid's intersections. In order to adjust the brightness of a specific pixel, a current must be passed between two grid conductors.

Furthermore, to reduce the amount of power needed to regulate each pixel's brightness, an active matrix places a transistor at each pixel's intersection. In an active-matrix display, this means the current can be turned on and off frequently, increasing the refresh rate. Some passive matrix LCDs are dual-scanned, which means that the grid is scanned with current twice as fast as it was scanned once with the older technology. The active matrix, on the other hand, is a far more advanced method.

### 4.4.1 Advantages

1. Bright images are generated due to high peak intensity

2. Zero geometric distortion
3. Flat Screens
4. Lesser consumption of electricity and less heat production

### 4.4.2 Disadvantages

1. Require precision in the tracking of pixels for noise reduction
2. Limited viewing angle
3. Saturation and compression occur when the LCD's bright end of the intensity scale is overworked
4. Slower response times may lead to poorer quality of image

## 4.5 LED TV

LED TVs replace cathode ray tubes to incorporate light-emitting diodes (LEDs) for illumination of display rather than the CCFLs used in standard LCD televisions. As a result of their high visibility in direct sunlight, they can be used as outdoor store signs or advertisements. Additionally, they have seen widespread use in variable-message signs along highways and as destination signs on public transportation vehicles in recent years. LED displays are an alternative to conventional projection screens because they provide higher contrast ratios and can be used to create massive, seamless video streaming.

Edge-lit displays have LEDs only along their outer edges, while full-array displays illuminate their entire backs with the same technology. However, Sony, Samsung, Vizio, and LG all use edge lit technology, while Sharp and Toshiba use full array lit technology. Full-array LED TVs, which illuminate the entire screen with LEDs, are bulkier and more expensive than edge-lit LED TVs.

### 4.5.1 Advantages

1. Provides good brightness of the screen to improve the user experience
2. Removes the light flickering across the screen
3. Allows for longer display usage
4. Eases control and displays specific colour characteristics using a variety of hues
5. Easily portable because it is thin and light-weight

### 4.5.2 Disadvantages

1. Very expensive
2. Must use a good quality antenna

## 4.6 PLASMA TV

One of the alternates in TV is Plasma TVs in which tiny pockets of gas convert to plasma upon application of a voltage to them. The application of voltage makes mercury collide within the plasma, to emit ultraviolet (UV) rays. In turn, it passes through phosphor cells to create an image. Three phosphor cells (red, green, and blue) combine to form a colour. Each pixel in plasma TV is self-emissive and able to produce its own light. They are able to provide a stunning true black effect because they do not need an external source of light to power the display. But a glass panel is required for a plasma TV, which is why they are relatively large and have a lot of glares during the day.

### 4.6.1 Advantages

1. Less bulky than rear projection television
2. Large screen sizes can be manufactured commercially
3. Much greater viewing angle
4. Can be wall mounted

### 4.6.2 Disadvantages

1. A picture is not so clear under bright conditions
2. Higher energy consumption
3. Doesn't work well at high altitudes
4. Produce glare due to reflection

## 4.7 HDTV

One of the most promising technologies in TV is HDTV (High-Definition Television). HDTV provides high-quality video by utilisation of digital signals. As a result, interference caused by analogue parts in electrical currents is eliminated. HDTV employs a wider aspect ratio of 16:9 resulting in better image perception. The higher resolution of HDTV produces much finer images, with more detail and colour than previous formats. Additionally, watching HDTV feels more natural because the screen's viewing angle is similar to that of the human eye. Direct-view TVs, plasma TVs, rear- and front-projection TVs, and front-projection HDTVs are all examples of HDTV displays. However, an HDTV tuner is needed to enjoy the highest quality format (1080i).

### 4.7.1 Advantages

1. An increase in resolution improves picture quality.
2. The use of interlaced scanning improves the visual experience.
3. Audio quality is greatly improved because Dolby technology is used.
4. The HDMI allows for automatic configuration and bidirectional communication between video sources.

### 4.7.2 Disadvantages:

1. Authentication delays are a potential cause of blank screens. It's also possible for there to be issues with the screen flashing.
2. The cost of an HDMI cable is higher per metre than that of an analogue one.
3. Greater heat production and power usage compared to alternative technologies.
4. Life expectancy is comparatively low.

## 4.8 3-D TV

Through the use of special glasses, viewers of 3-D televisions are given the impression that the images they are viewing have depth in addition to height and width, just like real-world objects. It makes use of techniques like multi-view display, stereoscopic display, 2D-plus-depth, and any other type of 3D display. The majority of modern 3D TVs employ an active shutter 3D system or a polarised 3D system, and some are autostereoscopic, requiring no glasses. Furthermore, it works by creating two distinct moving images, one for each eye. Because the image seen by the right eye is blocked from view by the left eye, we perceive depth.

Watching 3D TV requires a high-definition (HD) set with either passive or active 3D technology. Polarization, or the production of light rays with different properties in different directions, is essential to passive 3D technology. Separate, polarised images are presented to each eye via the 3D TV's display and the viewer's polarised 3D glasses; the brain then combines these images to create the 3D effect.

Unlike passive 3D TVs, which display identical left and right images, active 3D TVs rapidly swap between the two images. In turn, the viewer uses powered or active 3D glasses that pick up infrared signals from the TV. Therefore, an individual's right eye will not be able to see content meant for the left eye, and vice versa, thanks to synchronisation between the screen and the glasses. The images change so rapidly, typically around 50 times each second, that they become a single entity in the mind of the observer.

The use of lenses is one approach to producing glasses-free 3D television, which are paper-thin plastic lenses which may be positioned in front of a TV screen to send slightly different images to the audience's right and left eyes. In order for the "lenticular 3D TV" to work, the viewer must maintain a specific distance and viewing angle from the screen. Thus, the technology is considered more suitable for portable devices like laptops and DVD players than for large-screen televisions.

Even though the concept of 3D television is still in its early stages, holography, which uses laser light to create a 3D image, is being explored by manufacturers. The concept is to project an image onto a "cloud" in the midst of a room, making it viewable from any direction without the need for special 3D glasses.

## 4.9 PROJECTION TV

Projection TVs are classified into two types: front projection and rear projection. Both techniques employ tiny devices capable of producing extremely finer images. These devices can be reflective, which means that light bounces off the device and picks up the image. In addition, they may have a transmissive quality, allowing the image to be picked up by light as it passes through the gadget. After the light captures the image, a magnification task is done by lenses, followed by projection onto the screen.

A front-projection system employs a projector and a separate screen to project images onto the screen's front surface. Although a HD monitor is ideal, any flat, white surface can be used for projection. It enables users to recreate the large-screen experience of a commercial movie theatre in their own homes. Furthermore, it is the least expensive alternative for big-screen video at home and is easy to set up. Front projectors, on the other hand, look best in a darkened room, much like a movie theatre. A dark room allows for the most contrast and sparkle in the image. Most projectors necessitate maintenance that is not applicable in case of flatscreen and standard televisions.

Rear-projection systems resemble standard televisions. TVs with built-in projectors are able to display content by projecting an image onto the rear of a screen. A rear projection TV is significantly less expensive than a comparable-sized plasma TV. Rear projection televisions used to be thick and boxy, taking up a lot of floor space but a majority of them are now less than 18" deep, thanks to the new digital light engines. Moreover, it tends to perform reasonably well in rooms with some ambient light, as long as you view them head-on at eye level. However, it has limited viewing angles and tends to reflect anything illuminated by ambient light in the room.

**Fig. 4.3:** Rear projection TV [2]

**Fig. 4.4:** Front projection TV [3]

## 4.10 DTH CONCEPT BLOCK DIAGRAM

India is one of the world's largest DTH service providers. Because of the large number of viewers in India, the demand for DTH is greater than in any other country. Some of the popular DTH providers in India include TATA Sky, Dish TV, and Airtel DTH and others.

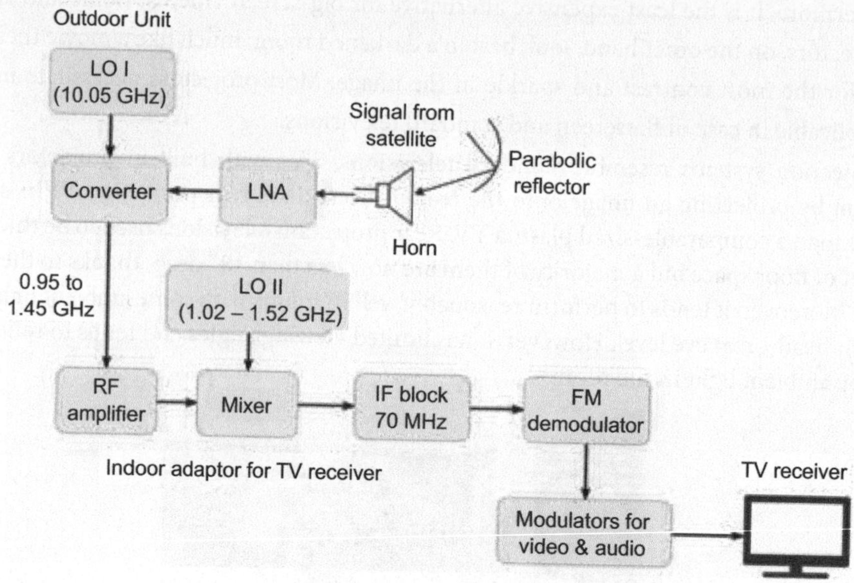


**Fig. 4.5:** Block diagram of DTH technology.

Using a small, personal dish for direct reception of satellite signals on a TV is referred to as DTH technology (Fig. 4.5). DTH services aimed at providing better quality satellite signals plus more channels to compete with local cable TV operators. The satellites compress and encrypt signals and then the transmission is done through high-powered geo satellites. Signals are received by dishes of DTH providers to the end users. This technology was traditionally known as Direct Broadcast Satellite (DBS) technology. The home receiver system of a DBS consists primarily of two units: an outdoor unit and an indoor unit.

### 4.10.1 The Outdoor Unit

The dish antenna at the receiver first receives the satellite's downlink signal, which ranges from 12.2 GHz to 12.7 GHz. The received signal is then directed to the receiving horn which directs the signal to the polarizer. This aids in the transmission of left-hand circular or right-hand circular polarised signals by performing good switching. This outdoor unit's schematic is depicted in Fig. 4.6.

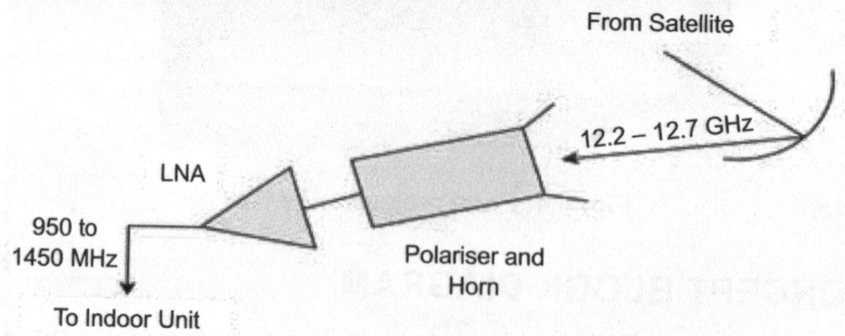


**Fig 4.6:** Schematic diagram of the Outdoor Unit

There is a low noise block, which is made up of a Low Noise Amplifier (LNA) and a downconverter. The LNA is the unit responsible for signal amplification at low levels because these are the signals that the antenna has difficulty recognising, and any necessary amplification must be performed without the addition of noise. Subsequently, the downconverter performs down-conversion of the signal in the range of 12.2 to 12.7 GHz, which is converted into the range of 950 to 1450 MHz. The down-converted frequencies are optimal for signal transmission from the connecting cable to the indoor unit.

### 4.10.2 The Indoor Unit

The tuner is the first component of the indoor unit (Fig. 4.7) and is responsible for selecting the transponder. The transponder maintains the guard band of 24 MHz in the selected

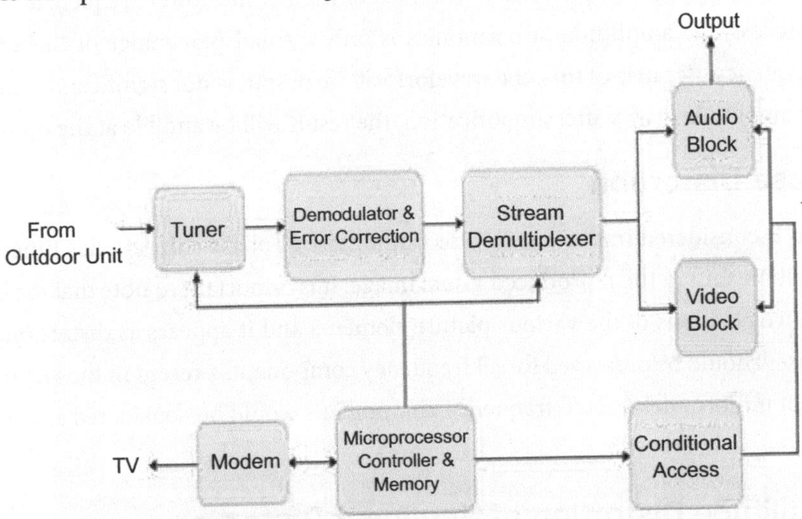


Fig 4.7: Block diagram of Indoor Unit

bandwidth. As a result, any of the 32 transponders must be received by the indoor unit. Only a signal from 16 transponders must be available for a single polarization and QPSK modulation of the carrier at the centre frequency. The quadrature phase-shift keying modulated signal is then demodulated and converted into the equivalent bitstream. After that, an error correction scheme is implemented to remove the errors from the received sequence.

## 4.11 VIDEO AMPLIFIER

The video amplifier must meet the following specifications to generate a favourable image on the picture tube's screen.

### 4.11.1 Video Gain

Video gain is referred to as the ability of the amplifier to produce a good raster for all values of intensity of the picture tube scanning beam.

### 4.11.2 Bandwidth

Bandwidth is referred to as the ability to reproduce a picture's all frequency range of information including Higher frequencies that are required for horizontal sections requirements.

### 4.11.3 Frequency Distortion

Frequency distortion is the change in the amplitude of harmonics after the amplification. The amplitude of the harmonics is a higher value than its effect. This normally occurs when there is a change in amplification level as a result of a change in frequency. An input signal to a practical amplifier will have both the fundamental frequency and harmonics of higher and lower frequencies. Because of the amplification process, the amplitude of harmonics is only a small percentage of the amplitude of the fundamental frequency. Because of this, the waveform at the output is not significantly altered. However, if the harmonic amplitude grows after amplification, the result will be audible at the output.

### 4.11.4 Phase Distortion

Phase distortion is considered important in video amplifiers as phase shift implies time shift, which in turn implies position shift in the reproduced visual image. It is important to note that the human eye can detect shift relative positions of the various picture elements and it appears as distortion. This requires that relative phases should be preserved for all frequency components present in the video. However, it is worth noting that uniform delay in all frequency components would be considered as would be to delay the entire signal.

### 4.11.5 Amplitude Distortion of Nonlinear Distortion

When the transistor's non-linear characteristics dampen the amplification applied to the wave's positive and negative extremes, a phenomenon known as amplitude distortion occurs. The wave's amplitude is reduced, but the parts of the waveform to either side of the centre are unaffected. However, the ear is much more sensitive to the associated change in the shape of the waveform tips, making the resulting decrease in amplitude in audio signals much less noticeable.

### 4.11.6 Manual Contrast Control

For better contrast values between the white and black portions of the image, there should be provided to control amplitude for the given video signal. Video amplifiers employ direct-coupled and RC-coupled configurations. Both configurations require high-frequency compensation.

## 4.12 SOUND SECTION

FM sound signal with a weak amplitude having a carrier frequency of 5.5 MHz, passes through a tuned amplifier, known as sound IF. This signal is then fed into the FM detector, which is typically either a ratio detection circuit or a discriminator circuit. Subsequently, the output is given to a limiter circuit. FM detectors generally have the following characteristics:

1. Obtained sound output signal is proportional to deviations from the sound carrier frequency.
2. Rate of frequency deviation is helpful to determine the frequency of the signal.

De-emphasis circuit at the output of the sound detector works as the counterpart of the pre-emphasis circuits (used at the transmitter) and restores the higher audio frequency amplitudes. Before being delivered to the audio output (power) amplifier, the audio signal is at least once amplified. Volume and tone controls are built into audio amplifiers and either single-ended or push-pull is used as a power amplifier. FM demodulator circuits are found in the form of integrated circuits (ICs) that drive one or two loudspeakers mounted at the receiver's front panel.

AGC circuit maintains controls over the gain of the RF and IF circuitries to nullify changes observed during a reception and to deliver nearly constant signal voltage to the video detector. a bias voltage generated in the AGC circuit changes the operating points of the amplifying devices. It is worth noting that regardless of the picture signal details, the sync levels of the composite video signal is unchanged and a rectifier circuit generates a control voltage in accordance with the sync level. The AGC control, as described above, is depicted by a block schematic circuit arrangement in Fig. 4.8.

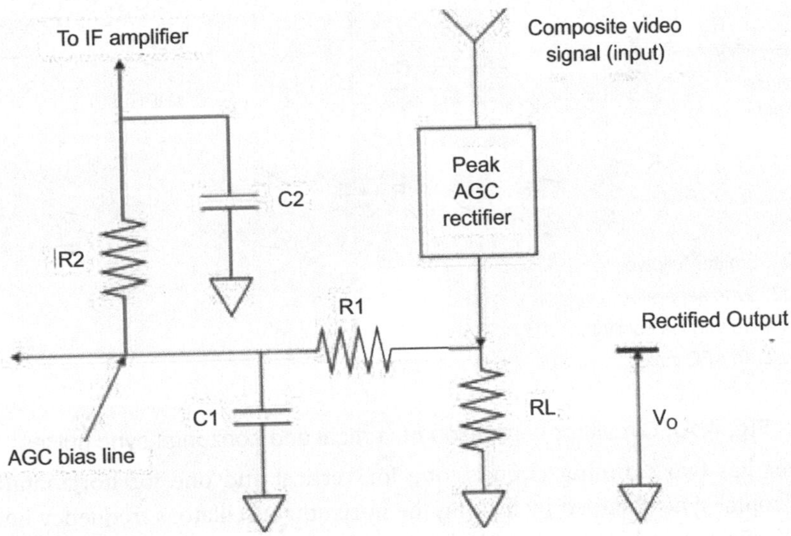


**Fig. 4.8:** Block schematic circuit arrangement of the AGC system

## 4.13 SYNC SEPARATION

The sync separator separates horizontal and vertical synchronising pulses from the video signal. Obtained sync output is simultaneously given to the horizontal and vertical deflection oscillators. Due to these signals, the image information can be correctly positioned on the raster screen. A block schematic diagram of the operation sequence is shown in Fig. 4.9.

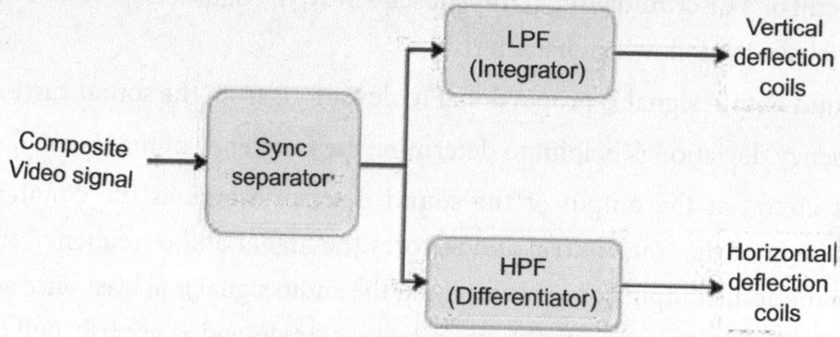


**Fig. 4.9:** Generalised Block diagram of sync separator

The sync separator works on the biasing of the circuit so that video signals cause current to flow in the device. Self-biasing is used as the dc bias voltage that is produced by the ac signal itself.

## 4.14 SYNC PROCESSING

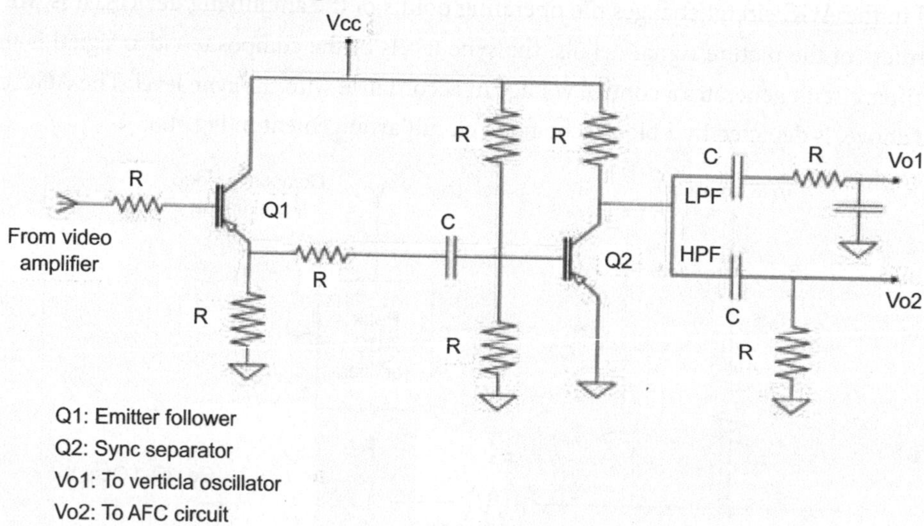


**Fig. 4.10:** Circuit for separation of vertical and horizontal sync pulses

The receiver has two scanning circuits, one for vertical and one for horizontal electron beam deflection. Horizontal sync obtained by locking the horizontal oscillator's frequency holds the picture's line structure. Vertical sync locks the vertical frames by triggering the vertical oscillator. Equalising pulses make even and odd fields vertically synchronised for good interlacing. Fig. 4.10 shows the circuit diagram used to separate out these sync pulses.

## 4.15 VIDEO INTERFACE

Video interfaces are used to transfer data from computer memory to a monitor for display. For the given line, input video signal voltage reproduces picture elements. The video signal's black level should cut off the grid voltage and the blacker-than-black grid voltage. At the picture tube grid, sync pulses are used to time vertical and horizontal scanning circuits in the receiver.

## 4.16 COMPOSITE VIDEO

Image data is not transmitted alone. Along with it, blanking and sync pulses are also transmitted. CVD (composite video signal) is the signal that contains all of these components and it can be represented with either positive or negative polarity. For positive polarity, the brighter the screen requirements, the greater the amplitude of the video signal available. The blanking level is zero, and the sync pulse is less than zero. As a result, the sync top is at its lowest point. On the other hand, for negative polarity, the brighter the screen, the smaller the amplitude. As a result, the sync pulse is positive, above the blanking level.

An RCA or "yellow plug" cable is a composite video cable, which has been used for transmitting a video signal. However, it cannot support HD content. Video signals range all the way from high definition to the lowest common denominator, composite. When compared to other video formats, the quality of video produced by a composite signal (which transmits all colour and brightness data over a single cable) is noticeably lower. Moreover, the composite video input and output is a standard feature on almost all video equipment. Therefore, with this setup, just one RCA or BNC plug is required. As video signals have to travel through a distance, there can be a significant loss of resolution and clarity of the picture. Losses due to radio frequency interference further may reduce picture quality. Common uses for composite video signals in professional editing settings include menu outputs, troubleshooting, and low-quality preview monitoring. They're also frequently used to connect electronics like VCRs and DVD players to home televisions.

## 4.17 COMPONENT VIDEO

Component video divides video contents into three cables — green, blue, and red — each of which transmits a different component of the video signal. The green cable (also known as the Y cable) transmits the signal's brightness-related information, the blue and red cables carry the blue and red components of the picture's colour and a combination of all three signals infers gray components. Because component video is distributed across three separate cables it can satisfy the need for HD resolutions to produce smoother images with more defined edges.

Component cables have now been used as video cables due to their promising high picture quality and HD support capabilities. Almost all modern HDTVs, most video game consoles, DVD players, digital video recorders, and cable set-top boxes also support component video.

## 4.18 SEPARATE VIDEO

Separate video (S-Video), is a video signalling standard with resolutions of 480i or 576i. This is a video analogue signal that contains information in two parts (a) chroma signal which stands for colour information and (b) luma signal which stands for brightness, contrast etc. It transmits video information in a single cable without combining it with audio signals. In comparison to composite video, S-Video can achieve better image quality but lower colour resolution. S-video signals are typically transmitted via a cable with a 4-pin mini-DIN connector.

## 4.19 DIGITAL VIDEO

DVI, or Digital Visual Interface, allows for the transmission of HD digital or analogue signals from computers or HD video devices to flat-panel screens or projectors. Moreover, up to 24 digital signals and 4 analogue signals can be transmitted via a DVI connector. Recently, all-digital workflows involving image capturing to projections have been in digital form for the following reasons:

Digital video is very effective for all the steps in communications including recording, storing, processing, manipulating, transmitting, receiving and recovering. being SNR very high it is almost error-free. According to the need and availability of bandwidth, compression and decompression on digital video signals are possible. Today digital communication has been widespread due to cheaper ICs, availability of advanced communication networks capable of high-speed communication, huge storage space availability, and advanced computing architectures digital video can now be handled at a variety of applications demanding data rates on mobiles, tablets and other networks.

## 4.20 SERIAL DIGITAL INTERFACE (SDI)

SDI was created to make it simple to convert traditional analogue component video to and from a digital video. Nowadays, the compression of digital video has become easy due to the development of SDI. The goal of the Serial Data Transport Interface (SDTI) is to solve this problem by transforming SDI into an interface that can transmit many different types of data without breaking backwards compatibility with current SDI router networks.

HD-SDI is a variant of SDI with a greater data rate to accommodate the transmission of uncompressed HD video. Similar to SDI, HD-SDI can transmit both audio and video over a single cable. Moreover, by combining two HD-SDI channels, certain devices can offer even greater data rates. This is also referred to as dual-link HD-SDI.

## 4.21 HIGH-DEFINITION MULTIMEDIA INTERFACE (HDMI)

HDMI transmits digital video data in uncompressed form while digital audio data into compressed or uncompressed form. Such a transfer requires an HDMI-compliant source device at the transmitter side plus a compatible monitor, video projector, or digital TV at the receiver side. HDMI digitalizes analogue video and it implements EIA/CEA-861 standards that specify video, audio and auxiliary data formats. Recent HDMI versions have optional advanced features such as 3D, and Ethernet extensions. Moreover, statistics show that approximately ten billion HDMI devices have been sold by January 2021.

## 4.22 DIGITAL VIDEO INTERFACE (DVI)

DVI is an abbreviation for "Digital Video Interface". The Digital Display Working Group (DDWG) developed the DVI video connection standard The majority of DVI ports can connect to both analogue and digital displays. DVI will convert the digital signal into an analogue signal if the display requirement arises. However, there is no need for conversion if the display is digital.

DVI connections are classified into three types: DVI-A is for analogue, DVI-D is for digital, and DVI-I is for integrated (both analogue and digital). DVI can support signals with a bandwidth in excess of 160 MHz, and hence can be used for high-resolution displays like UXGA and HDTV.

## 4.23 CD PLAYER

A compact disc (CD) is a plastic disc used to store data such as video, audio, and text files. The signal to be recorded on CD is amplified and then transformed into a digital signal using a sample and hold circuit and an ADC. The ADC output is also used by the Laser Beam Generator and the signal from the crystal oscillator and Laser beam generator controls both the control circuit and the servo system. The servo system, which is powered by a motor, controls the rotation of the disc as well as the track and focus of the Laser beam generator. After recording, the unexposed photoresist material is chemically removed, leaving a helical pattern across the surface of the glass disc. This is used to create the glass master for mass-production CDs. CD recording system block diagram is shown in Fig. 4.11.

### 4.23.1 Working

Within the CD player is a tiny laser beam (a semiconductor diode laser) and an electronic light detector (photoelectric cell). Pressing the play button activates an electric motor that spins the disc at a rapid speed. From the CD's centre to its outer edges, the laser beam is activated and scanned along a track by the photocell. Along the scanning process, the motor gradually slows the disc's rotation. This is because the disk's actual surface moves faster than the laser and photocell, requiring progressively more time to read the same amount of information as the disk's distance from the centre increases.

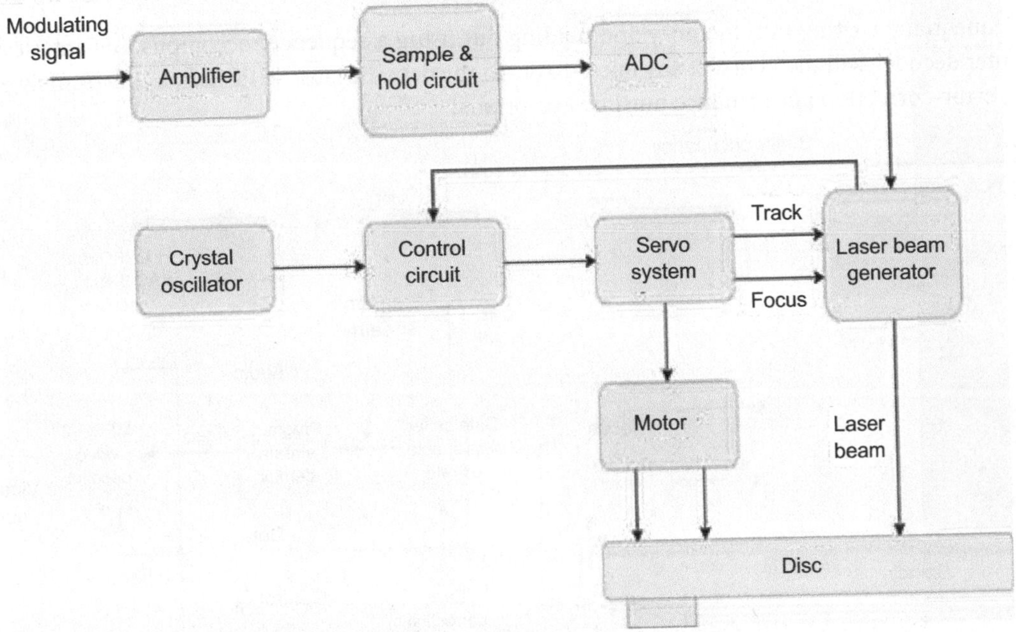


**Fig. 4.11:** Block Diagram of CD Recording

The laser beam is shone upward onto the CD's shiny underside, where it is reflected by the disc's pits and lands. Unlike the pits, which scatter the laser light, the lands reflect it straight back. The reflected light is detected by the photocell a surge of current is sent to the electronic circuit which generates the number one. When the photocell cannot detect any light, as a result of an absence of land, the electronic circuit returns a value of 0.

So, the sequence of binary digits (0s and 1s) recorded on the disc at the manufacturer is gradually reconstructed by the scanning laser and electronic circuit. The DAC then takes these binary numbers and translates them back into a progressing electric current pattern. In the end, the electric currents are converted into audible sounds by a loudspeaker.

## 4.24 DIGITAL VERSATILE DISC / DIGITAL VIDEO DISC (DVD) PLAYER

Fig. 4.12 shows the basic components of a DVD player. The block diagram divides into control/servo and data paths. The control part of the system gives user connectivity with the servo mechanisms and ensures the safety and proper operation sequence.

A servo system includes the focus and tracking servos, as well as a motorised loading drawer and chucking mechanism. The most common form of player that uses power loading is one that features a drawer to store the disc in the issue. A safety switch prevents the laser from turning on inadvertently if the machine's drawer or lid is opened while it's functioning. The goal is to ensure that the user takes no risk at all.

In order for the MPEG decoder to be able to decode the video, the data path must include the data separator along with a de-interleaving and error-correcting mechanism as well as a RAM buffer. Digital data is generated from the waveform sent from the EFM plus readout via the data separator. EFM plus coding is a group code, so any error that alters the channel patterns can lead to the loss of up to eight bits of data.

Sequentially writing into memory and reading out using a sequencer accomplish de-interleaving. The outer decoder will then correct any burst errors in the data. Because MPEG data is extremely error-prone, error-correction performance must be exceptional.

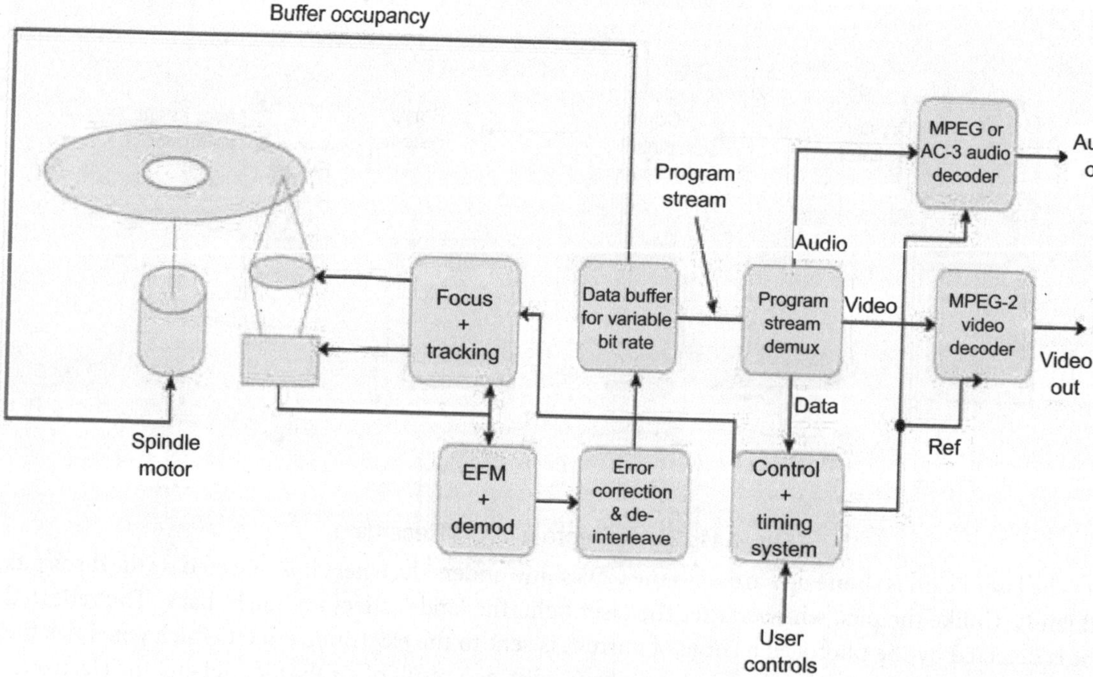


**Fig. 4.12:** Essential components of a DVD player.

A MPEG programme stream emerges after the de-interleave and outer error-correction processes. Some of the data in the programme stream will be video, while others will be audio, and this will be routed to the appropriate decoder. The bit rate of this programme stream is not fixed in DVD but can vary depending on the difficulty of the programme material to maintain consistent image quality. The bit rate is altered by varying the disc speed.

## UNIT SUMMARY

Simple PAL colour system is modified using a delay line (PAL-D) to compute averaging before presenting the colour to the human eye. The receiver of PAL-D Colour TV is divided into various sections namely tuner, sound strip, AGC, sync-separator & deflection circuits, luminance channel, colour signal processing and subcarrier generation and control. The evolution of television has resulted in the invention of Digital, LCD, LED, Plasma, HDTV, 3-D TV, and Projection TV.

DTH technology refers to the reception of TV signals from satellites to a personal dish. A DBS's home receiver system is made up primarily of two units: an outdoor unit and an indoor unit. Moreover, the video amplifier, sound section, sync separator, and processor are all integral parts of the television receiver. A video amplifier is used to amplify, buffer, and filter analogue video signals over 75W cabling to maintain signal fidelity. Sync separator is used to separate synchronising pulses from the composite video signal and subsequently, TV receiver has two scanning circuits, one for vertical electron beam deflection and the other for horizontal electron beam deflection. Furthermore, data is transferred from computer memory to a monitor via video interfaces.

A composite video cable is a traditional video signal transmission standard that uses a single cable and connector. Component video divides video signals into three cables: green, blue, and red, each transmitting a different component of the video signal. S-Video is a 480i or 576i video signalling standard and this video analogue signal contains two signals: chroma and luma. Additionally, Serial Digital Interface (SDI) makes it easy to convert analogue component video to digital and HDMI transfers uncompressed video and digital audio from HDMI-compliant devices to compatible receivers. The digital video interface supports signals with more than 160 MHz bandwidth, allowing it to be used for UXGA and HDTV. A CD is a plastic disc used to store data such as video, audio, and text files, while a DVD stands for a digital versatile disc or digital video disc.

## EXERCISES

### Multiple Choice Questions

1. Which of the following specifications should the video amplifier meet in order to produce a favourable image?
   (a) Low frequencies
   (b) A constant amplitude of video signal
   (c) Weak video signal
   (d) The relative phases of the frequency components present in the video signal must be preserved.
2. Which of the following is not a component of a PAL-D colour television receiver?
   (a) Tuner
   (b) Demodulator
   (c) Luminance channel
   (d) Sync-separator

3. Which statement about Plasma TV is true?
   (a) Doesn't work well at high altitudes
   (b) Bulkier than rear projection television
   (c) Lesser energy consumption
   (d) Limited viewing angle
4. In the indoor unit of the DBS system, the transponder maintains the guard band of ______ in the selected bandwidth.
   (a) 24 Hz (b) 240 Hz (c) 24 MHz (d) 240 MHz
5. What are video interfaces used for?
   (a) To provide reception of satellite signals on a TV
   (b) To transfer data from computer memory to a monitor for display
   (c) To generate the correct frequency subcarrier
   (d) To generate audio output.

| ANSWERS | | | | | | | | | |
|---|---|---|---|---|---|---|---|---|---|
| 1. | (d) | 2. | (b) | 3. | (a) | 4. | (c) | 5. | (b) |

## Short and Long Answer Type Questions

1. State and briefly describe any 2 types of televisions.
2. What is the function of a luminance channel in the receiver of PAL-D Colour TV?
3. What is direct-to-home technology?
4. Why has digital video become more popular in recent years?
5. Discuss composite video and component video.
6. Describe the working of a CD player with the help of a suitable diagram.
7. Write a short note on the video amplifier.
8. Explain the PAL-D colour TV receiver and draw the block diagram.
9. Explain the sound section of a television receiver with the help of a suitable diagram.
10. Explain sync separation and processing in detail.

## PRACTICAL

1. To study sync separate section of TV receiver
2. To study the video amplifier section of TV receiver
3. To study AGC circuitry of TV
4. To study the working of Digital, LCD, LED, Plasma, HDTV, 3-D TV, and Projection TV and
5. To study SDI and HDMI interfaces of TV

## KNOW MORE

Business-to-business (B2B) and business-to-customer (B2C) interactions can both benefit greatly from the use of video. It has the potential to revolutionise a company's product offering and increase operational efficiencies. All of this activity has one overarching purpose: to increase sales. There has been a dramatic rise in the importance of online video in just the initial few months of 2020.

We have been engulfed by limitless video applications in our day-to-day lives because streaming media is now integrated into everything from vacuums to watches to medical equipment. Video-enabled health monitors, remote surgical capabilities, and smart hospitals outfitted with AI-based surveillance are essential in today's modern healthcare system. Additionally, visiting a store or the gym in person has largely been replaced by digital alternatives like shoppable video and digital fitness. Consequently, video accounts for 80% of all internet traffic, according to Cisco. With the increasing attention being paid to AR/VR by tech giants like Meta, the internet traffic is only predicted to increase.

The cloud has enabled flexibility and expandability more than ever before in video production, processing, delivery, and other areas. As a result, organizations can innovate faster and expand further with the cloud. Moreover, remote-friendly workflows would continue to benefit from virtualized video environments.

There is more content available than ever before. However, without a simple way to organise, store, and access this content, it is merely digital clutter. Artificial intelligence (AI) is now being used in video content management systems (CMS) solutions for content analysis, compilation, highlight production, indexing, searchability, and other forms of automation — all of which are game changers for content distributors with a large video library.

According to Cisco, by 2023, the number of internet-connected devices will exceed three times the global population. Universities, hospitals, and every other type of organisation would employ video to blur the line between in-person and remote experiences. The evolution of this field has always astounded us and would continue to do so in the future.

A group of engineers from Sony and Philips collaborated in 1979 to develop the first digital audio disc. In 1980, the Red Book CD-DA standard was published following a year of testing and debate. Moreover, after their 1982 commercial debut, compact discs and players quickly gained widespread acceptance.

In the 1920s, 16mm projectors were used to create the first home theatres and in the 1930s, technological advances allowed for the creation of 8mm and sound 16mm film.

Once the cost of a Kodak 8 mm film projector dropped, showing home movies became increasingly common among middle- and upper-class American households in the 1950s.

Surround sound was first used in a film in 1940 for the Disney animated short Fantasia. The operatic composition Flight of the Bumblebee by Nikolai Rimsky-Korsakov inspired Walt Disney to include a bumblebee in his musical Fantasia and make it seem as if it were flying in all corners of the theatre.

When the REDD 17 stereo mixing system was built at London's Abbey Road Studios by EMI's Record Engineering Development Department in 1958, it marked a significant step forward in the recording industry.

From 1911 through 1915, a series of experiments were conducted in which a microphone and loudspeaker were linked to a 12-volt battery, resulting in the first such instance of auditory feedback. Further studies led to the creation of 'Magnavox,' the first electric public address system.

## REFERENCES

1. LCD TV, accessed 29 August 2022, <https://www.indiamart.com/proddetail/lcd-tv-13675533891.html>
2. Rear projection TV, accessed 29 August 2022, <https://www.researchgate.net/figure/Rear-projection-TV-with-UHP-lamp-RCA_fig8_238963512 >
3. Front projection TV, accessed 29 August 2022, <https://hdguru.com/category/sxrd/>
4. Gulati, R. R. (2005). Monochrome and colour television. New Age International.
5. Gupta, R., Nanda, S. J., & Panchal, P. (2018). A colour transformation approach to retrieve cloudy pixels in daytime satellite images. In Proceedings of the Second International Conference on Computational Intelligence and Informatics (pp. 311-319). Springer, Singapore.
6. Watkinson, J. (2012). Introduction to digital video. Routledge.
7. How Stuff Works, How Digital Television Works, accessed 29 August 2022, <https://electronics.howstuffworks.com/dtv.htm>
8. Display Mate, LCD Advantages and Disadvantages, accessed 29 August 2022, <https://www.displaymate.com/lcds.html>.
9. Yale Tools, 9 Advantages and Disadvantages of LED TV, accessed 29 August 2022, <https://yaletools.com/advantages-and-disadvantages-of-led-tv/>.
10. Camponotes.com, Advantage and Disadvantages of Plasma Display Screens, accessed 29 August 2022, <http://camponotes.blogspot.com/2014/11/advantage-and-disadvantages-of-plasma.html>.
11. Britannica, HDTV broadcasting, accessed 29 August 2022, <https://www.britannica.com/technology/HDTV>
12. Wikipedia, 3D television, accessed 29 August 2022, <https://en.wikipedia.org/wiki/3D_television>
13. How Stuff Works, How Projection Television Works, accessed 29 August 2022, <https://electronics.howstuffworks.com/projection-tv.htm>
14. BrainKart.com, Direct to home Broadcast (DTH): Block Diagram, Advantage, accessed 29 August 2022, <https://www.brainkart.com/article/Direct-to-home-Broadcast-(DTH)--Block-Diagram,-Advantage_11962/>
15. Sims, H. V. (1969). Principles of PAL Colour Television and Related Systems. Iliffe.
16. Hutson, G. H. (1971). Colour Television Theory: PAL-system Principles and Receiver Circuitry. McGraw-Hill Companies.
17. Fink, D. G. (1957). Television engineering handbook. McGraw-Hill.
18. Grob, B. (1975). Basic Television, principles and servicing. McGraw-Hill Companies.

# 5 Home/Office Appliances

## UNIT SPECIFICS

Through this unit we have discussed the following aspects:

- Introduction, advantages and detailed Working of the microwave oven;
- Detailed Working, types and specifications of a washing machine;
- Role of microcontrollers in Home or office appliances
- Brief about the application of a fuzzy system for Home or office appliances applications

The practical applications of the topics are discussed for generating further curiosity and creativity and improving problem-solving capacity.

Besides giving a large number of multiple-choice questions as well as questions of short and long answer types marked in two categories following the lower and higher order of Bloom's taxonomy, assignments through a number of numerical problems, a list of references and suggested readings are given in the unit so that one can go through them for practice.

After the related practical, based on the content, there is a "Know More" section. This section has been carefully designed so that the supplementary information provided in this part becomes beneficial for the users of the book. This section mainly highlights the initial activity, examples of some interesting facts, analogy, history of the development of the subject focusing the salient observations and finding, timelines starting from the development of the concerned topics up to the recent time, applications of the subject matter for our day-to-day real life or/and industrial applications on variety of aspects, case study related to environmental, sustainability, social and ethical issues whichever applicable, and finally inquisitiveness and curiosity topics of the unit.

## RATIONALE

Nowadays human daily routines are full of utilising modern electronic equipment. In household or office usage of equipment like microwave oven and washing machine are very common. Understanding of benefits, and drawbacks along with operating instructions is needed for all common citizens. At the same time, in order to make them available to common people and maintenance knowledge of basic operating principles, components, and technical specifications is essential for technicians and designers. In addition to the above, in this module understanding the controller, fuzzy logic, and technical specifications of a washing machine have all been discussed.

## PRE-REQUISITES

Mathematics: Electronics Circuits (Class XII)
Physics: Signal and Wave (Class XII)

## UNIT OUTCOMES

List of outcomes of this unit are: After learning this unit student will be able to

**U5-O1:** Describe working principles of microwave oven and washing machine

**U5-O2:** Use consumer electronics devices with ease at home or offices

**U5-O3:** Identify and Diagnose primary working faults of consumer electronics devices

**U5-O4:** Explain concepts of controllers, fuzzy logic, and technical specifications in the context of normal usage

| Unit-5 Outcomes | EXPECTED MAPPING WITH COURSE OUTCOMES (1- Weak Correlation; 2- Medium correlation; 3- Strong Correlation) | | | | | |
|---|---|---|---|---|---|---|
| | CO-1 | CO-2 | CO-3 | CO-4 | CO-5 | CO-6 |
| U5-O1 | 1 | 1 | - | - | 3 | 3 |
| U5-O2 | 3 | 2 | 2 | 2 | 2 | 3 |
| U5-O3 | 3 | 2 | 2 | 2 | 3 | 2 |
| U5-O4 | 2 | 2 | 2 | 2 | 2 | 2 |

## 5.1 MICROWAVE OVEN

### 5.1.1 Introduction

It was by chance that microwave ovens were first discovered. An American electrical engineer named Percy Spencer was experimenting with a magnetron in the 1950s to help aircraft and ships navigate in bad weather or darkness. One day, though, he happened to have a bar of chocolate in his pocket when the magnetron turned on, and he was astounded to find that the bar melted almost immediately from the intense heat. Because of this, the concept of using a magnetron to cook food entered his mind. Popcorn was the "proof of concept" that led him to the idea of creating a universal microwave oven.

**Fig. 5.1:** Microwave oven [1]

A microwave oven (electronic oven) is a small, box-shaped appliance that rapidly heats food by exposing it to a strong electromagnetic field at a high frequency. An electron tube called a magnetron is responsible for producing the radiation used in microwave ovens, which has a frequency of about 2,450 megahertz. Because of the vibrations caused by microwaves, molecules like water, fat, and sugar (among others) become heated. As a result, baking and other cooking tasks that take several hours in a traditional oven can be finished in minutes in a microwave oven because the heating happens inside the food rather than warming the surrounding air. However, they can cause foods to be cooked at a different rate or unevenly due to the absorption heating process. In addition, microwave ovens can't be used to cook food in metal containers because the metal inhibits microwaves.

## 5.1.2 Precautions, Advantages & Disadvantages

Users must adhere strictly to all safety instructions and guidelines in their microwave oven manuals. A few examples of such safety measures are:

1. Use only microwave-appropriate cooking utensils.
2. If the microwave's door is distorted, bent, or otherwise damaged, do not use it.
3. No one, especially children, should ever stand directly in front of a running oven for a prolonged period of time.
4. Consistently wipe the oven's interior, edge, and door with water and mild detergent.

### 5.1.2.1 Advantages

1. The food can be cooked in a short time duration while retaining its natural flavours and nutritional value.
2. Frozen food can be defrosted quickly.
3. Cooking with a microwave reduces the need for oil, enabling you to make low-fat meals.

### 5.1.2.2 Disadvantages

1. It is not suitable for deep frying.
2. Using a microwave poses risks to health because it has been linked to the production of carcinogenic agents. Moreover, reduced immunity causes pathological alterations in the human blood.
3. As compared to traditional stovetop cooking, microwaves alter food's texture, colour, and flavour.

## 5.1.3 Working

Due to the microwave's ability to focus heat on individual molecules, food can be cooked in a short period of time. This results in dielectric heating, whereby the rotation of the food's polar molecules generates heat. Because excitation is relatively uniform in the outer 25-38 mm (1-1.5 inches) of a homogeneous (high water content) food item, microwave ovens can rapidly and efficiently heat food without drying it out.

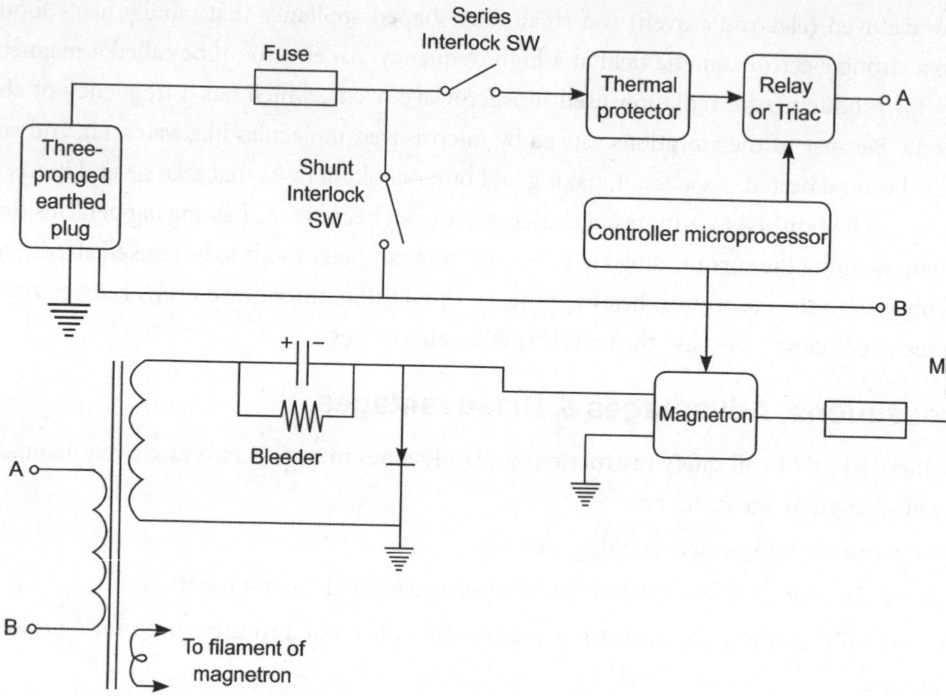


**Fig. 5.2:** Diagram of a microwave oven

Microwave radiation is a type of electromagnetic radiation that does not cause ionisation and is used to heat food in a microwave oven. A three-pronged, earthed plug or socket is required for use with a standard household electrical outlet. When the oven door is shut, power will be supplied to the mains transformer. Because there is at least one interlock switch in series with the primary transformer, not even a speck of dust on the relay or trial can cause the oven to turn on if the door is ajar. Moreover, dielectric heating occurs when water, fat, and other substances in the food absorb microwave energy. As a result, molecules (including those of water) rotate as they try to align themselves with the alternating electric field of the microwaves because they are electric dipoles. This implies that they have a partial positive charge at one end and a partial negative charge at the other and when molecules rotate at high speeds, they collide with one another and set them in motion, releasing energy in the process. Spread out as molecular vibration in solids and liquids, this energy is quite useful. Fig. 5.2 is a diagrammatic representation of a microwave oven's internal components and a brief explanation of the major components is given in next sub-section.

## 5.1.4 Components

Here are a few of the most crucial parts of any microwave:

1. **High voltage transformer**: A high-voltage step-up transformer is installed within a microwave oven because its power requirements exceed those of the standard voltage carried by a home's electrical wiring. 230- 240 V supply is boosted to very high voltage levels and then fed into the cavity magnetron.

2. **Cavity magnetron**: A cavity magnetron is a high-powered vacuum tube. It can transform electrical energy into long-range microwave radiation.
3. **Microcontroller**: It is a device that facilitates two-way interaction between humans and electronic devices. This controlling component incorporates a central processing unit (CPU), along with other elements such as RAM and I/O circuitry, to perform various tasks. Furthermore, it interprets the user's commands and displays them on the microwave's screen. A microcontroller inside the microwave receives commands from a keypad, displays the results on an LCD screen, and activates or deactivates the microwave generator via a relay.
4. **Waveguide**: The waves from the magnetron's output are directed into the cavity by a waveguide, which is a metallic tube with a hollow interior where the food is placed.
5. **Cooling fan**: The magnetron's efficiency and lifespan are both improved by the use of cooling fans, which keeps the device from overheating during operation.

### 5.1.5 Technical Specifications

**Table 5.1:** Technical specifications of a commonly used microwave oven

| Supply voltage | Single phase alternating current, 220V, 50Hz |
|---|---|
| Power Consumption | ~ 1300W |
| Microwave power | 0.7 - 0.85 kW |
| Microwave Frequency | 2.45 GHz |
| Oven Capacity | 10 - 25 litres |
| Accessories | A glass or ceramic tray is usually included. |

## 5.2 WASHING MACHINE

### 5.2.1 Introduction

The technology behind washing clothes has come a long way over the centuries, from gruelling manual labour to sophisticated machine cleaning. Domestic washing machines were operated manually until the late 1800s, while commercial models were powered by steam and belts. In 1908, however, the Thor, the very first commercial electric washer, changed all of that. Furthermore, the United States saw the introduction of the earliest electrical clothes dryers in the years leading up to World War One.

Now it is possible to do laundry without ever having to check on the washing machine, thanks to the convenience of automated settings. You can wash your clothes in an automated fashion without having to keep an eye on the washing machine. Moreover, technological advancements have made it so that washing machines can take in the necessary water and detergent and set a timer for cleaning, rinsing, and drying depending on the mode and the number of garments.

### 5.2.2 Classification based on Washers

There are two distinct types of washers, the top-loading variety and the front-loading variety, which differ only in the direction in which the clothes are loaded.

**Fig. 5.3:** Front loading washing machine [2]

To further categorise the top-loading washers, there are both semi-automatic and fully-automatic models. These are further described in detail below:

1. **Front loading washing machine**: The garments are loaded into the front of this machine. Moreover, research shows that front-loading washers are more efficient than top-loading ones, using less water and detergent while still yielding clean clothes.
2. **Top-loading washing machine:** Clothes are loaded and unloaded from the top of this fully automatic washing machine, which features a round vessel that serves as the washer, rinser, and dryer. In many regions, it has surpassed the popularity of the front-loading model.
   (a) **Semi-automatic washing machine:** Both the washer and the dryer use their own dedicated tubs, and their individual timers allow for precise control over the cycles. Clothes can be washed by placing them in a wash vessel, filling them with water, adding detergent, and then setting a timer. Then the washer will automatically shut off after the allotted time has passed and the clothes can be collected and dried in the sun, or a partial drying cycle can be run in the dryer vessel.

Fig. 5.4: Semi-automatic top load washing machine [3]

(b) **Fully automatic machine**: In a fully automatic model, the same tub is used for all three processes—washing, rinsing and drying. The number of clothes is detected by the machine's sensors, and it then calculates the appropriate volume of detergent and water to use, as well as the optimal wash and dry times.

Fig. 5.5: Fully automatic top load machine [4]

### 5.2.3 Working

The System Controller, which is the brain of the system, of a washing machine regulates the flow of electricity to the machine's many sensors, motors, and pumps. Additionally, it powers the screen that demonstrates the progress of the washing. Several parts of a washing machine, depicted in Fig. 5.6, have been described below.

1. **Display panel**: Touch screens allow users to manage every aspect of a machine from a single interface.
2. **Sensor**: A low-water-level alert is sounded by a water-level sensor in the washing machine. Moreover, when the washer detects that all of the clothes have been cleaned, it will emit a beep and allow you to open the door.
3. **Driving Motor**: Both "reverse" and "forward" rotations are possible with a bidirectional motor. It turns because the current is flowing in a forward direction, which is also the direction of the motor's rotation. Thus, the driver in the opposite direction does the opposite.
4. **System Controller**: The system controller monitors the motor's speed and adjusts it accordingly during various processes. This also works for the maintenance of any sensor, be it a door or pressure sensor, keypad, or speed sensor.

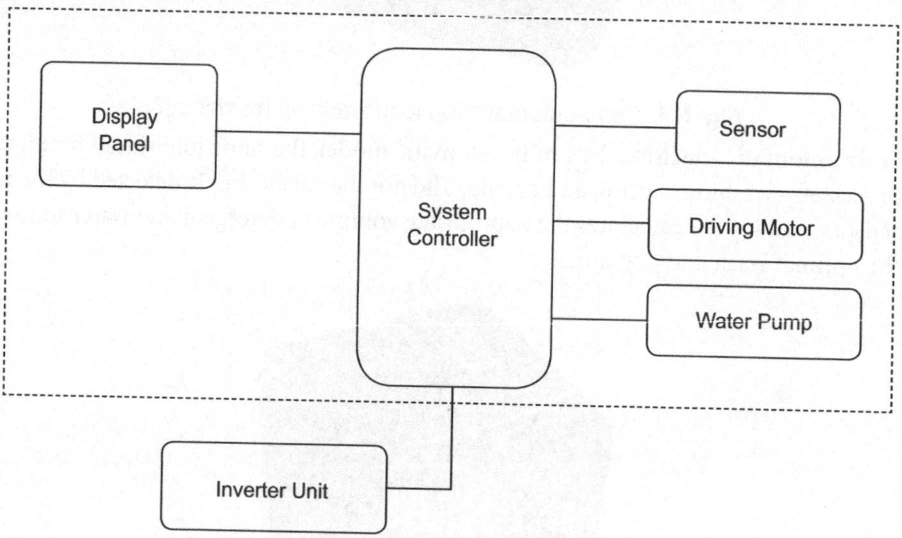


**Fig. 5.6:** Block diagram of a washing machine.

5. **Water Pump**: In order to recycle clean water and get rid of dirty one, a water pump is required. The drain line is connected to the pump's bottom half, while the pump's upper half recycles the rinse water. The pump is powered by an electric motor, which can switch directions. When the washer is in the wash cycle, the water is recirculated, and the drum spins in one direction; in the spin cycle, the water is drained, and the drum spins in the opposite direction.

## 5.2.4 Wiring Diagram of Washing Machine

The wiring diagram, connection steps, and internal circuit of an electric washer are detailed in the following Fig. 5.7.

1. **Spin motor**: The clothes are dried with a Spin Motor, also called a dryer motor. They are single-phase induction motors with two windings: a starting winding and a running winding. This type of spin motor typically makes use of a capacitor to turn on. The washing machine's motor has a higher power rating than the spin motor, hence the spin cycle takes so long.
2. **Wash motor**: To clean the fabric, a wash motor is used which is faster and has more torque than the spin motor. Moreover, to move wet clothes, it needs to be more powerful. It is a single-phase induction motor with two windings, much like a spin motor, and it needs a capacitor to get going. The wash timer controls the rotation of the wash motor in both directions.
3. **Spin timer**: The spin motor can be controlled with the aid of the spin timer. It has two connections, one to the spin motor and the other to the power source. It is important to note that both semi-automatic and fully-automatic washing machines utilise the spin motor and spin timer.
4. **Wash timer:** The wash timer can be set to run for a certain amount of time, reverse the direction of the washing machine's motor, and sound an alarm or buzzer when the cycle is finished. Wash timers can have anywhere from three to six terminals, and there are a variety of different types.
5. **Door switch:** The door switch enables the use of the washer with the door closed and no other conditions met. Since the door switch is wired in series with the spin motor, the latter won't turn on unless the door is closed.

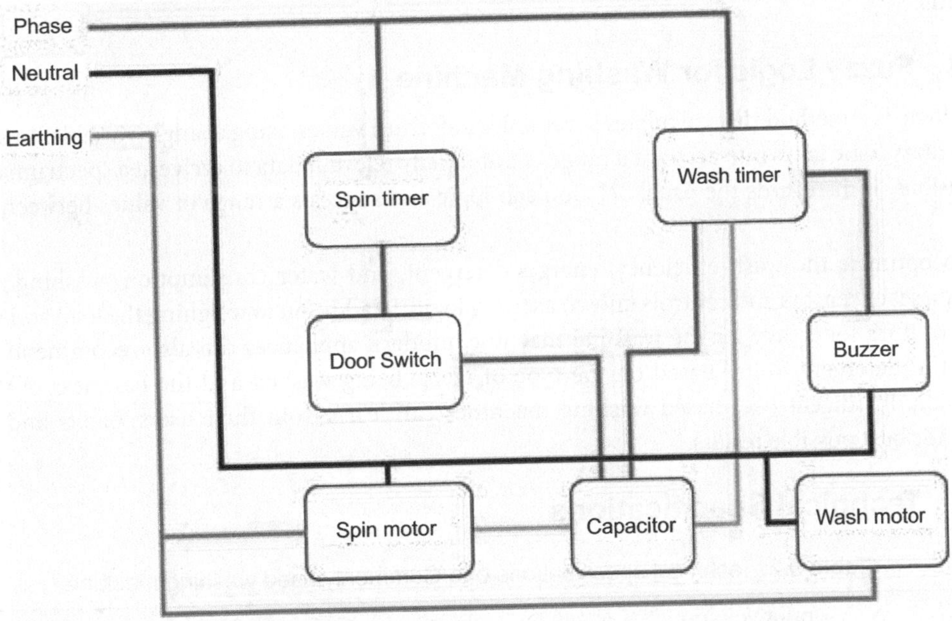


**Fig. 5.7:** Connection diagram and wiring of a washing machine.

### 5.2.5 Controller of Washing Machine

Herein is presented an effective electronic replacement for the mechanical controller of the washing machine's single-phase motor (Fig. 5.8).

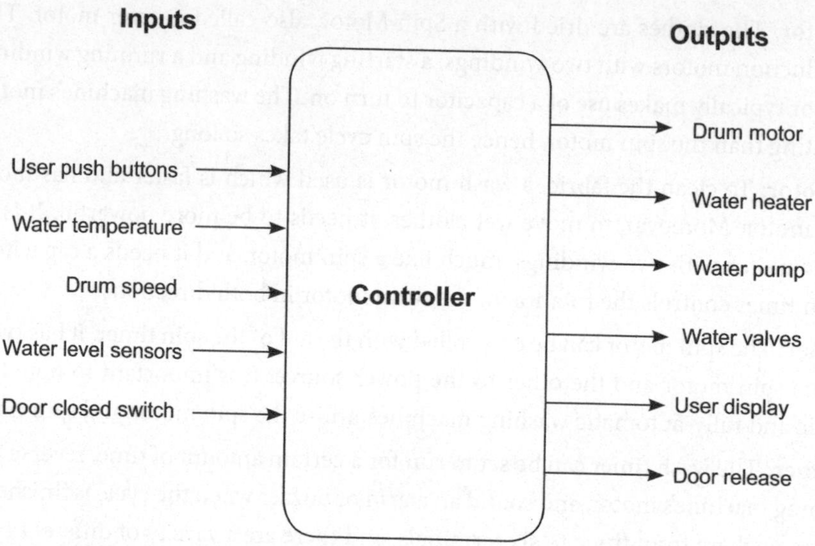


**Fig. 5.8:** Inputs and outputs of a washing machine controller.

A single-phase motor needs a master timer to determine how long the motor should run (washing time), and a spin direction controller to reverse the direction of rotation every 10 seconds.

### 5.2.6 Fuzzy Logic for Washing Machine

Fuzzy logic is a method that calculates several distinct truth values using a single variable. In simplified terms, fuzzy logic takes into account a range of potentially relevant data to arrive at a spectrum of correct conclusions. It disregards the usual 0/1 Boolean logic and suggests a range of values between true and false.

To optimise the wash efficiency, energy, detergent, and water consumption, washing machines employ a set of sensors and controls known as fuzzy logic. In addition to weighing the load and sounding an alarm if it's too heavy for the washing machine, modern appliances can also recommend the ideal amount of detergent to use based on the type of fabric being washed and the hardness of the water. Artificially Intelligent (AI) based washing machines can learn from their users' habits and adapt to deliver the best possible results.

### 5.2.7 Technical Specifications

**Table 5.2:** Technical specifications of a Commonly used washing machine

| | |
|---|---|
| Input voltage | 100V - 240V |
| Input power | 1.2 kW |
| Output power | 0.1kW - 0.4 kW |
| Capacity range | 3kg - 18kg |
| Maximum efficiency | Approx. 31% |

## UNIT SUMMARY

Almost very modern home would have a microwave oven and a washing machine. Due to the microwave oven's extraordinary ability to rapidly heat, defrost, and cook food, this accidental discovery has quickly gained widespread attention. This device uses a powerful electromagnetic field at a high frequency to perform its functions; its core components are a high voltage transformer, a cavity magnetron, a microcontroller, a waveguide, and a cooling fan. Moreover, its power consumption is around 1.3kW and it employs a microwave frequency of 2.45 GHz.

The innovations in laundry care over the years have been remarkable. The automated settings of today's washing machines make it possible to do laundry without constantly checking on the machine. Top-loading and front-loading washers are two different kinds of washing machines that vary only in the direction in which one loads laundry. Top-loading washers can be further broken down into two groups: the semi-automatic and the fully automatic. There is a system controller integrated with a *display panel, sensor, driving motor, and water pump* that all work together to accomplish the task of washing. Furthermore, fuzzy logic is a set of sensors and controls used in washing machines to improve the machine's performance and reduce its use of resources like water, energy, and detergent. The typical washing machine has an efficiency of 31% and an output power of 0.1 kW to 0.4 kW.

## EXERCISES

### Multiple Choice Questions

1. Which of the following is not a disadvantage of using a microwave oven?
   (a) It cannot be used for deep frying.
   (b) Frozen food cannot be defrosted.
   (c) It may reduce immunity.
   (d) It might change the taste of the food.
2. Typically a microwave oven has a frequency of ____
   (a) 2450 MHz (b) 2.45 MHz (c) 1300 MHz (d) 1.3 MHz
3. Which of the following is a component of a washing machine?
   (a) Waveguide (b) Analogue to digital converter
   (c) Generator (d) System controller
4. ______ can be set to run for a certain amount of time and reverse the direction of the washing machine's motor.
   (a) Wash motor (b) Wash timer
   (c) Door switch (d) Spin timer
5. Which of the following is not an input of controller in a washing machine?
   (a) Water level sensors (b) Water heater
   (c) Water temperature (d) Drum speed

| ANSWERS | | | | | | | | | |
|---|---|---|---|---|---|---|---|---|---|
| 1. | (b) | 2. | (a) | 3. | (d) | 4. | (b) | 5. | (b) |

## Short and Long Answer Type Questions

1. State the advantages and disadvantages of a microwave oven along with the precautions that the user must take.
2. What are the technical specifications of a microwave oven?
3. Write a short note on a washing machine.
4. State the technical specifications of a washing machine.
5. Draw the controller of the washing machine.
6. Describe the working principle of a microwave oven and draw a suitable block diagram.
7. State and describe the integral components of a microwave oven.
8. Which components of a washing machine aid in its operation?
9. Explain the wiring diagram of a washing machine with the help of an appropriate diagram.
10. Describe the different types of washing machines and state their technical specifications.

## PRACTICAL

1. Study the working principle of the washing machine and its technical specifications.
2. Study the working of microwave.
3. Program(s) to implement various controlling /measurement applications in home appliance using controllers

## KNOW MORE

- Since the first microwave oven was built, which stood at about 1.5 meters (5 ft) in height, the technology has shrunk considerably, and millions of units have been sold around the world.
- Due to the dipole nature of water molecules, microwave heating is optimal, while it is less so for fatty and sugary substances.
- Microwaves operating at 2.45 GHz are typically found in residential ovens, whereas 915 MHz microwaves are more typical in commercial and industrial kitchens.
- The Romans developed a primitive soap that they called "lye" which was used to wash their clothes. This soap was made out of animal fat and ash. Moreover, garments were traditionally washed in a big pot or cauldron, boiled, and then laid on a flat board and beaten with a paddle in colonial times.
- French inventors called their early 1800s washing machine invention the ventilator. A metal barrel drum perforated with holes served as the device, which was rotated manually over a fire.
- As early as 1960, Lotfi Zadeh from the University of California first addressed the idea of fuzzy logic. He was investigating challenges involved in using computers to comprehend human speech.

## REFERENCES

1. Microwave oven, accessed 25 September 2022, <https://www.zilan.com.tr/microwave-ovens/microwave-oven-1-1>
2. Front loading washing machine, accessed 25 September 2022, <https://www.bosch-home.in/productlist/washer-dryer/front-loading-washing-machines/WGA244ASIN#/Togglebox=manuals/>
3. Semi-automatic top load washing machine, accessed 25 September 2022, <https://www.voltasbeko.com/washing-machines/7-kg-semi-automatic-washing-machine-wtt70dt.html>
4. Fully automatic top load machine, accessed 25 September 2022, <https://www.reliancedigital.in/kelvinator-8-kg-top-loading-fully-automatic-washing-machine-with-error-monitoring-and-memory-backup-kwt-a800sg/p/491604437>
5. ThoughtCo., A Brief History of Washing Machines, accessed 27 September 2022, <https://www.thoughtco.com/history-of-washing-machines-1992666>
6. Bright Hub Engineering, What is a Washing Machine? Types of Washing Machine, accessed 27 September 2022, <https://www.brighthubengineering.com/consumer-appliances-electronics/38358-types-of-washing-machines/>
7. ETechnoG, Washing Machine Wiring, Internal Circuit and Connection Diagram, accessed 27 September 2022, <https://www.etechnog.com/2021/11/washing-machine-wiring-connection-circuit.html>
8. US Food & Drug Administration, Microwave Oven Radiation, accessed 26 September 2022, <https://www.fda.gov/radiation-emitting-products/resources-you-radiation-emitting-products/microwave-oven-radiation>
9. Britannica, Microwave Oven, accessed 25 September 2022, <https://www.britannica.com/technology/microwave-oven>
10. Studious Guy, Microwave Oven Working Principle, accessed 26 September 2022, <https://studiousguy.com/microwave-oven-working-principle/>
11. Explain That Stuff, Microwave Ovens, accessed 27 September 2022, https://www.explainthatstuff.com/microwaveovens.html

❑❑

# 6 Digital Electronic Systems

## UNIT SPECIFICS

Through this unit we have discussed the following aspects:

- Fax understanding;
- *Operating Principle of Photocopies;*
- Simple workflow of Air Conditioner;
- Understanding of Working Principal of Refrigerators;
- Working of Digital Camera;
- Basic functionality of Camcorder

The practical applications of the topics are discussed for generating further curiosity and creativity and improving problem-solving capacity.

Besides giving a large number of multiple-choice questions as well as questions of short and long answer types marked in two categories following the lower and higher order of Bloom's taxonomy, assignments through a number of numerical problems, a list of references and suggested readings are given in the unit so that one can go through them for practice.

After the related practical, based on the content, there is a "Know More" section. This section has been carefully designed so that the supplementary information provided in this part becomes beneficial for the users of the book. This section mainly highlights the initial activity, examples of some interesting facts, analogy, history of the development of the subject focusing the salient observations and finding, timelines starting from the development of the concerned topics up to the recent time, applications of the subject matter for our day-to-day real life or/ and industrial applications on variety of aspects, case study related to environmental, sustainability, social and ethical issues whichever applicable, and finally inquisitiveness and curiosity topics of the unit.

## RATIONALE

Various digital electronic devices including a fax machine, photocopier, air conditioner, refrigerator, digital camera, and camcorder are discussed in this module. Understanding of basic operational principles and fault diagnosis is required in order to enhance the life span of these products. This module explores the fax machine's details, its origins, functionalities, operation, and the basics of sending and receiving faxes. Additionally, the components and operating principles of the photocopier, also known as electrophotography, have been covered. One of the most widespread types of technology, air conditioners come in a wide variety of models. Moreover, aspects of AC operation are also discussed in this unit. A number of different parts, including the refrigerant fluid, compressor, condenser, thermostatic expansion valves, evaporator, and receiver, carry out a refrigerator's cooling process. Digital cameras, including their operation and classification, have also been outlined. Camcorder fundamentals have been highlighted in the last section of the chapter.

## PRE-REQUISITES

Mathematics: Electronics Circuits (Class XII)

Physics: Signal and Wave (Class XII)

## UNIT OUTCOMES

Outcomes of this unit are : After learning this unit student will be able to

**U6-O1:** Describe the basic operation of FAX

**U6-O2:** Describe the components of the Camera

**U6-O3:** Explain the working of Refrigerators

**U6-O4:** Realize the role of camcorders

**U6-O5:** Apply the knowledge of Digital Cameras to solve problems

| Unit-6 Outcomes | EXPECTED MAPPING WITH COURSE OUTCOMES (1- Weak Correlation; 2- Medium correlation; 3- Strong Correlation) | | | | | |
|---|---|---|---|---|---|---|
| | CO-1 | CO-2 | CO-3 | CO-4 | CO-5 | CO-6 |
| U6-O1 | - | - | - | - | 3 | 2 |
| U6-O2 | - | - | 3 | 2 | 3 | 3 |
| U6-O3 | - | - | - | - | 3 | 3 |
| U6-O4 | - | - | 3 | 2 | 3 | 2 |
| U6-O5 | - | - | 2 | 2 | 3 | 2 |

## 6.1 FAX

### 6.1.1 Introduction

Although the concept of a fax machine dates back to the 1800s, the widespread use of fax machines did not occur until the 1980s. A fax machine (also known as a facsimile machine) employs a conventional telephone line to transmit and receive documents. In order to send or receive a fax, an image must first be digitised by converting it into a grid of dots. It is possible to tell whether a dot is "on" or "off" based on whether it is black or white. Moreover, each dot represents a 1 (on) or a 0 (off) in a digital system. The receiving machine decodes the information from binary code to dots to reprint the image.

Fax machine also has an optical scanner to convert paper documents into digital format, a printer to physically deliver incoming faxes, and a phone line to connect the two entities. However, it is widely known that the resolution of an optical scanner is lower than that of a dedicated standalone scanner. Fax machines that use thermal paper have thermal printers, which use heat to print faxes.

Fax systems can be constructed by integrating a fax modem and an optical scanner, rather than purchasing an independent fax machine. You might not need the optical scanner if the documents which is be sent are digital. A fax machine has different functionalities, such as:

1. **Speed**: Fax modems can send and receive data at rates ranging from 4,800 to 28,800 bits per second. Depending on the fax machine, one page can take anywhere from 10 to 20 seconds to send data at 9,600 bps rate.
2. **Printer type**: The thermal printer used in most fax machines requires particular paper, which quickly turns yellow or brown with use. A common feature of higher-end fax machines is the ability to print on standard bond paper.
3. **Paper size**: Most fax machines support thermal paper in either an 8.5-inch or 10.1-inch width. When using the device, narrow-sized paper is imperative.
4. **Paper cutter**: Since thermal paper typically comes on a roll, the majority of fax machines include a paper cutter. However, few economical models or portable faxes may not have it.
5. **Paper feed:** If you need to send multiple pages at once, most fax machines have paper feeds that make it convenient.
6. **AutoDialing**: In order to send and receive faxes, users can choose from some different dialling options. To take advantage of cheaper phone rates, you can schedule the fax to send the document at a later time.

**Fig. 6.1:** Fax machine [1]

### 6.1.2 Working

To put it simply, a regular fax machine combines the functions of a telephone and a photocopier. To send a fax, the user first loads the necessary paperwork into the machine's document feeder before dialling the phone number of the intended recipient's fax machine. The original document is pulled by a gear mechanism over an optical scanner, and the scanner records the document's light and dark areas as a series of dots in rows or columns. Then the dots are read by a photoelectric cell and converted into electronic impulses that are sent down the phone lines to the recipient's fax machine. Fig. 6.2 depicts a block diagram of a typical fax network.

Electrical impulses are converted into dots by the receiving fax machine. The fax machine's internal print mechanism receives the decoded signal and prints a replica of the original document. Additionally, fax machines all over the world are compatible with each other because of international standards.

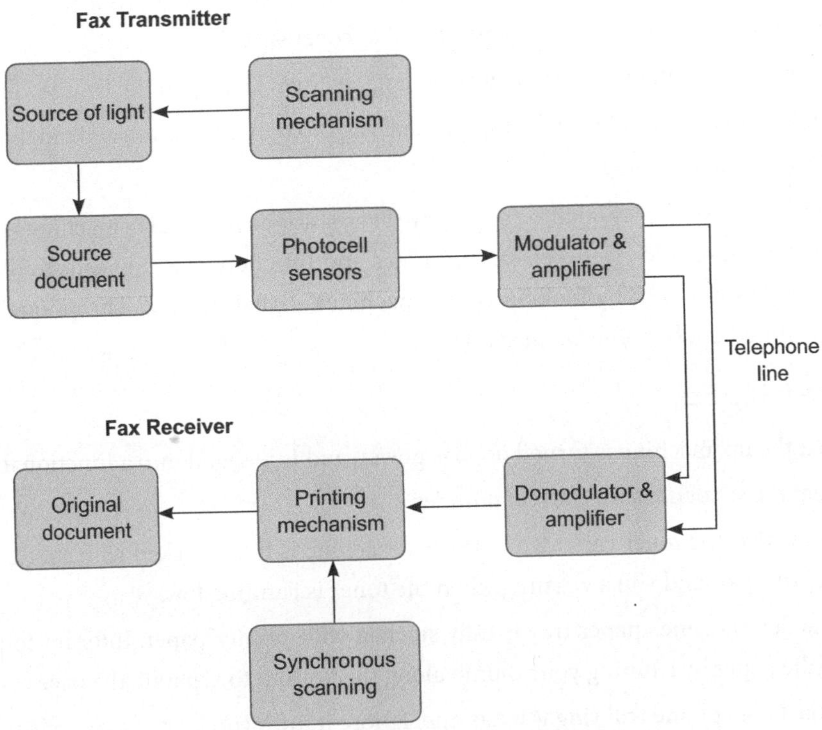


**Fig. 6.2:** Block diagram of a fax system

Similar to how a photocopier would photograph an original document in order to create a digital copy, fax machines scan individual pages and create digital images. The images must be modulated into a sound that can be recognised by another computer or fax machine to be transmitted over the telephone line. To send a fax, the sending machine makes a series of audio tones, which are then decoded by the receiving machine as characters. Thereby, demodulating the data into a digital image that corresponds to the scanned image broadcast by the scanner. Furthermore, the receiving machine will acknowledge receipt of the entire fax once the transmitting machine indicates the last page has been sent. Some fax machines will print a confirmation page after sending to confirm that the fax was sent successfully.

### 6.1.3 Operating Principle of Fax Machine

Despite the widespread adoption of electronic mail and web-based faxing services, familiarity with the operation of the traditional fax machine remains a valuable workplace skill. The following is a summary of the primary concepts involved in sending and receiving faxes.

#### 6.1.3.1 Sending Fax

1. Ensure the fax machine is powered and connected to a working phone jack.
2. Activate the fax machine.
3. Get the number of the receiving fax machine.
4. Compile all the documents to verify that they are received in the correct order.
5. Include the recipient's name, fax number/phone number, your name, your phone number, and a brief message on a separate piece of paper called a "cover sheet."
6. Place the cover sheet on top of the documents and lay them face up in the fax machine's feeder tray.
7. Call the fax recipient's number (take care for dialling instructions for international calls).
8. Hit the "fax" or "send" button.

A series of fax tones will play once the pages have been scanned into the fax machine's memory. These tones facilitate communication between fax machines and function as a "handshake." Verify whether your fax was successfully sent by looking at the machine's digital display. An optional confirmation report may be printed out by some fax machines.

#### 6.1.3.2 Receiving Fax

1. Check that the fax machine is turned on, has power, and is plugged into a functional phone jack.
2. Ensure that the sender has the correct number.
3. Check to see that the toner cartridge in the fax machine is fully stocked with ink. Lately, most fax machines will also send you a warning when the toner is running low.
4. Ensure the fax machine's paper tray is fully stocked with printer paper. In order to prevent paper jams, fan the paper by running your thumb along the bottom to separate the pages.
5. The fax machine's phone will ring if it has one. Ignore it until you hear a series of beeps that sound like a "handshake" to know that your fax machine is communicating with the sender.
6. After the first page is printed, the fax machine will start printing the rest of the fax automatically.
7. Double-check the cover sheet to ensure you got the right number of pages.
8. Office etiquette dictates that you confirm receipt of a fax by calling or emailing the sender, especially if it is a particularly important document.

## 6.2 PHOTOCOPIER

### 6.2.1 Introduction

In 1938, Chester Carlson established the photocopier, which he called electrophotography, and after 10 years, in 1948, his invention was officially recognised and renamed xerography. The Xerox Corporation began operations in 1949 and introduced the world's first plain paper copier.

**Fig. 6.3:** Photocopier [2]

Photocopiers rely on the law of attraction and the property of some materials to increase their electrical conductivity when exposed to electromagnetic radiation for their core operating principle. Printing on paper requires the use of toner, a negatively charged powder. At the very centre of a photocopier is a drum that is positively charged by static electricity. The master copy is laser-transferred onto the drum. The image's white parts (like paper's blank spaces) become more negatively charged as their charge is drawn away by the light, while the image's black components (like the text) retain their positive charge. Because it is attracted to positive surfaces, toner accumulates in the drum's darkest areas. The drum is what draws the different colours of toner (cyan, magenta, yellow, and black) when making colour copies. Countless hues are possible when these four prominent colours are mixed and blended.

## 6.2.2 Features

The integral components required for the photocopying procedure are as follows:

1. **Photoreceptor drum (or belt):** It is an important component that looks like a metal roller coated with photoconductive material, which is typically a semiconductor such as germanium, selenium, or silicon. This layer becomes an insulator when it is not exposed to light, blocking the flow of electrons between atoms. However, when the layer is illuminated, the photon energy relieves the electrons and allows current to flow. It is the newly liberated electrons that cancel out the positive charge that coats the drum to create the latent image.
2. **Corona wires**: A corona wire will create a field of positive charges on the drum and the paper. Applying a high voltage to these wires causes static electricity, which is then transmitted to the drum and paper.
3. **Lamp and lenses:** To make a photocopy, a powerful light source is needed to eject electrons from photoconductive atoms. The energy in the green and blue sections of the visible spectrum is sufficient to power this procedure. UV light has enough energy to produce a photocopy, but it is so intense that it can cause significant harm to our eyes and skin. Because of this, a regular

incandescent or fluorescent bulb is used to shine light onto the original document in photocopiers. Moreover, the lamp moves along the inside of the machine, lighting up a single strip of paper at a time as the machine feeds it in. Light from the lamp assembly is reflected from a mirror and then focused by a lens onto a drum that spins below. Because of the lens, you can target an exact area to concentrate a duplicate of the image.

4. **Toner:** Dry ink is another name for the fine, negatively charged powder that is toner. The toner in a toner cartridge is stored on larger, positively charged beads. As the drum is rolled over with toner-coated beads, the toner particles are drawn to the positively charged ions on the drum's surface where they have not been exposed. Electrostatic attraction between the paper and the particles increases as time passes, and upon heating, the plastic particles liquefy and permanently bond the colour to the paper.
5. **Fuser:** The fuser completes the process of fusing the toner onto the paper, making the image permanent. The fuser does two things: it melts the toner and presses it into the paper without letting either the toner or the paper stick to the fuser. These goals are achieved by employing quartz tube lamps and Teflon-coated rollers.

### 6.2.3 Working

To begin, the cylinder is electrostatically charged using a high-voltage wire (a corona wire or charge roller). The original paper is then illuminated by a bright light source, and the drum's white areas are copied in photoelectric manner. The light-exposed portions of the drum become conductive and discharge to the ground. A negative charge is maintained in the dark regions of the drum, which correspond to the black in the original document. Due to its positive charge, toner is attracted to and adheres to the drum's negatively charged areas during the imaging process (black areas). Subsequently, the drum's toner image is transferred to a piece of paper with a relatively stronger negative charge. Ultimately, the toner is liquefied and fused to the paper using rollers that apply heat and pressure.

Furthermore, creating a negative photocopy reverses the original document's colours, so that white text is superimposed on a black background. There are cases where the clarity and conciseness of a document are improved by making a negative photocopy of an old or faded original.

## 6.3 AIR-CONDITIONING

### 6.3.1 Introduction

It is commonly known as A/C or AC, air conditioning is the practice of reducing the temperature and, in some cases, the humidity inside a building or other enclosed area in order to make its occupants more comfortable. The following are a few types of air-conditioning units:

1. **Mini-split:** Ductless systems (also known as mini-splits) are frequently used for multi-zone or multi-split applications, which permits up to eight individual rooms to be conditioned by their indoor unit in conjunction with the outdoor unit. However, the length of the refrigerant lines connecting the outdoor unit to the indoor units is the primary challenge of multi-split systems.

2. **Ducted central systems:** The condenser is the external component of a split-system central air conditioner, while the fan coil unit (FCU), air handling unit (AHU), or evaporator is located indoors and serves to transfer heat to the outside air. Ductwork runs from the FCU to the rooms that need cooling.
3. **Central plant cooling:** Due to air's low density and heat capacity, large central cooling plants may use transitional coolants such as chilled water pumped into air handlers or fan coil units adjacent to or within the spaces to be conditioned, which then duct or deliver cold air into the areas to be conditioned. In order to cool the water, chillers in the plant use a refrigeration system, and the excess heat is usually released into the atmosphere through cooling towers.
4. **Portable units:** The indoor component of a portable system is on wheels and is linked to the outdoor component through adaptable pipes. There are air ducts that lead outside for the exhaust of hose systems, which may be monoblock or air-to-air. The water is collected in a bucket or tray of the monoblock variety, which automatically turns off when it is full. Continuous operation is possible for air-to-air types because they recycle the water through evaporation before discharging it through the ducted hose. However, the cooling efficiency of portable air conditioners is reduced because they draw air from inside and release it outside via a single duct.
5. **Packaged air conditioner**: Packaged air conditioners (self-contained units) are centralised systems that incorporate all the components of a split central system into a single housing, and then distribute the cooled air to the desired rooms, typically through ductwork. Units can be ground-based or rooftop-based (called "rooftop units"), cooled by water, refrigerant, or air, and used to condition either indoor or outdoor air. The air conditioning system in an outdoor unit is typically cooled by a cooling tower, while the air conditioning system in an indoor unit is cooled by liquid.

## 6.3.2 Working

The vapour-compression cycle, used in conventional air conditioning systems, transfers heat through the forced circulation and phase change of a refrigerant between gas and liquid to achieve cooling. While an air conditioner and an air source heat pump have similar parts, the latter has a reversing valve that allows it to warm and also cool a building. If the evaporator coil surface is relatively cooler than the dew point of the surrounding air, the absolute humidity of the air of the system would get decrease. To maintain a comfortable level of human activity, an HVAC system installed in an area will maintain a relative humidity of 30-60%. Fig. 6.4 illustrates the cooling cycle in which a refrigerant is processed in the following steps:

1. A cold liquid refrigerant is circulated through an evaporator, where it draws heat from the warmer room and dissipates it.
2. Once in its gaseous state, the refrigerant is compressed to raise its temperature.
3. The condenser coils do the work of dissipating the refrigerant's heat into the ambient air.

4. Then the refrigerant expands in order to lower its pressure and cool down to a temperature lower than the ambient temperature before the cycle can begin again.

Furthermore, for maximum water condensation, the evaporator temperature is lowered during the dehumidification cycle of most modern air conditioners by running the compressor while slowing the fan. In situations when the external air is cooler than the internal air, it is not necessary to use a compressor. Consequently, free cooling is preferred to ensure high efficiency in cooling.

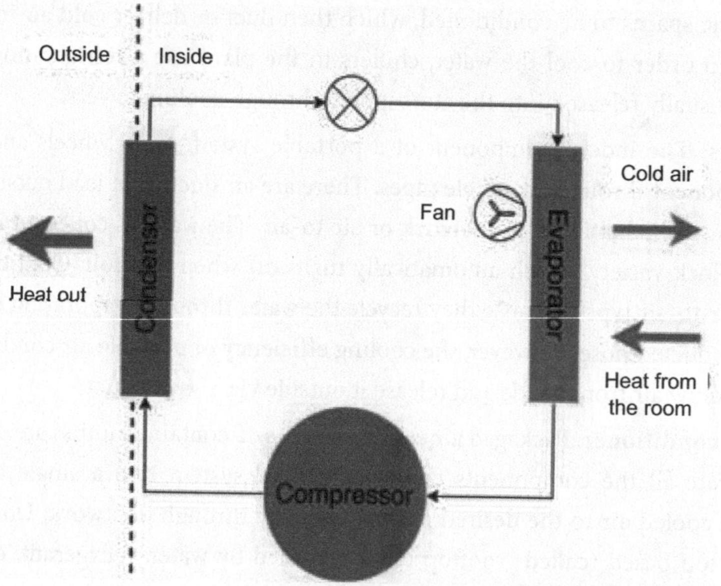


**Fig. 6.4:** The cooling cycle of an air-conditioner

Reverse-cycle air conditioners generate heat by switching to air source heat pump mode and reversing the refrigeration cycle. Heat pumps are more proficient than electric resistance heating because they transfer energy from the surrounding air or groundwater to the heated space. The heat pump's indoor evaporator coil is converted into a condenser coil and used to generate heat when the system is in heating mode and the outdoor condenser unit acts as an evaporator, expelling cold air.

## 6.4 REFRIGERATOR

### 6.4.1 Introduction

The refrigerator is the most common item in almost every household nowadays. Refrigerators make our lives easier in many ways, from preserving our food by reducing bacterial activity to providing us with cold beverages to enjoy during the hot summer. The modern refrigeration method can be traced back to the work of American inventors Oliver Evans and Jacob Perkins in the 19th century. Perkins created the first refrigeration system based on the ideas of Evans, who developed the vapour-compression refrigeration cycle. Even though Perkin's system was not widely popular, it did pave the way for the refrigerators we use today.

**Fig. 6.5:** Refrigerator [3]

### 6.4.2 Components of a Refrigerator

Some of the parts of a modern refrigerator which help in the cooling process are described below:

1. **Refrigerant Fluid**: A refrigerant fluid, which can be either a gas or liquid, circulates throughout a refrigerator and helps to regulate the temperature inside by changing phases to facilitate the conversion of thermal energy into mechanical energy. For example, hydrochlorofluorocarbon chemicals, more commonly known as freons, are among the most widely used refrigerant fluids, and they all have very low boiling points.
2. **Compressor**: The compressor, which is the mechanical "heart" of the refrigeration system, decreases the volume of gas in order to increase its pressure. The job is to mechanically raise the temperature of the refrigerant liquid by directing it through a narrower pipe. The gaseous refrigerant enters the compressor at low pressure and low temperature and exits at high pressure and high temperature. Additionally, it serves as an electric pump to circulate the refrigerant, and it is typically found in the appliance's lower back.
3. **Condenser**: The condenser, a mesh of pipes, dissipates the refrigerator's internal heat. Since the heat is dissipated through the condensation of the refrigerant fluid from a vapour to a liquid state within the condensing coils, the device has been aptly named a condenser. Condensation forms when the temperature drops below the condensation point, and this process releases heat.
4. **Thermostatic Expansion Valves**: A thermostatic expansion valve regulates the flow of liquid refrigerant into the evaporator. A thermostat activates or deactivates the compressor to regulate the cooling process. A refrigerator's compressor is automatically shut off by a sensor once the temperature inside has been determined to be optimal. It monitors internal temperatures and, if necessary, restarts the cooling process by turning on the compressor.

5. **Evaporator**: It is responsible for producing cold air inside the appliance, which is accomplished by drawing heat from within the refrigeration system so that it can be dissipated through the condenser.
6. **Receiver**: The receiver, also known as the liquid receiver, is a pressurised container used to store the liquid refrigerant. A receiver stores liquid refrigerant and supplies a steady supply to the expansion device.

### 6.4.3 Working

In accordance with the second law of thermodynamics, "heat cannot pass from a cold to a hot body without some other change, connected therewith, occurring at the same time," a refrigerator can maintain a constant temperature by utilising this principle. Consequently, mechanical work needs to be performed on the system to transfer heat from an area with a cooler temperature to an area with a higher temperature. The refrigeration system performs the necessary mechanical work. The refrigeration cycle (Fig. 6.6), also known as the vapour-compression refrigeration cycle, is a cyclical process that moves heat from a cooler to a warmer area.

When a refrigerator's compressor turns on, it forces a volatile refrigerant liquid through the evaporator's cooling coils and into the refrigerator's interior. Because of the latent heat provided by the food items, the refrigerant evaporates inside these coils, changing its phase from liquid to gas, thereby lowering the temperature inside the refrigerator. Next, the vapours are sent through an electrically powered compressor, which increases the pressure and, thus, the temperature of the vapours. These extremely hot and pressurised vapours are then directed toward the condenser coil, where they once again undergo a liquefaction process, this time changing their phase from vapour to liquid. This second phase transition allows the heat to escape into the surrounding environment. After passing through an expansion valve to lower its pressure, the liquid refrigerant is stored in a receiver or liquid reservoir until the next cycle begins. This process is repeated until the set temperature is achieved. In order to further boost the efficiency of the evaporator and condenser coils, some high-efficiency refrigerators use fans.

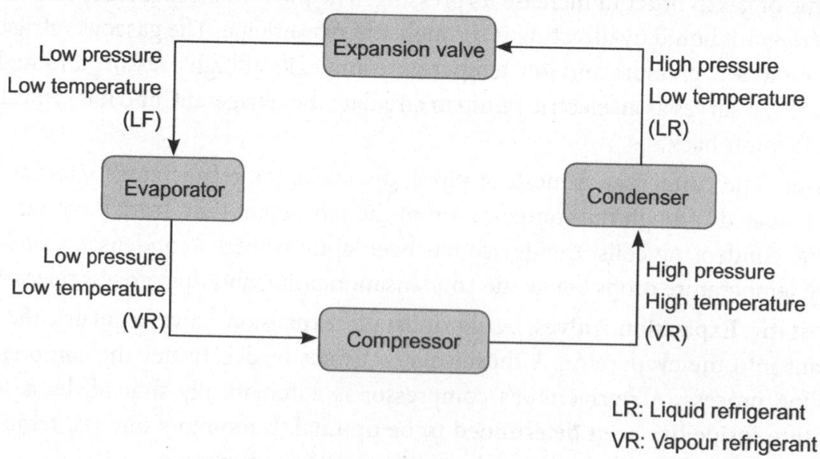


**Fig. 6.6:** The refrigeration cycle.

## 6.5 DIGITAL CAMERA

### 6.5.1 Introduction

The Greek word "Kamara" is the source of the modern English word "camera." Cameras are optical devices capable of capturing and storing visual data or transmitting it to a remote location. Those pictures could be still shots or videos, like movies or videos. There are many similarities between the human eye and the modern camera, which descended from the camera obscura.

The past three decades have seen tremendous changes in photography thanks to the introduction of digital technology. The optical principles behind a digital camera are the same as those behind a film camera, with the addition of a tiny photosensitive chip. A second chip or memory card then stores the electric signals as digital data after the first chip has converted the light energy into electric signals. This information can be interacted with on desktop computers, uploaded to various social media platforms, and printed out and pasted into traditional photo albums.

A digital camera works by focusing light onto an image sensor, which is made up of millions of small pixels. Through the photoelectric effect, photons are absorbed by each pixel on a CCD (Charge Coupled Device) or CMOS and electrons are released. For each photon that strikes the sensor, a charge is produced. This voltage is then "stepped up" by an amplifier before being sent to an analogue-to-digital converter (ADC).

### 6.5.2 Types of Cameras

1. **Point and Shoot cameras**: The most elementary variety is the point-and-click digital camera, which only requires you to aim it and press a button to take a picture.

**Fig. 6.7:** Point and Shoot camera [4]

The camera's ability to perform all necessary calculations and adjustments for taking a picture makes it a good choice for those who lack the time, interest, or talent to study photography. Smartphone cameras are one application of this technology.

1. **Twin-Lens Reflex (TLR) cameras**: It's a camera configuration where both lenses have the same focal length. The "taking lens," or photographic objective, is the component responsible for actually capturing an image, while the "viewfinder lens" is used to frame the scene for the photographer from above at waist level.

Fig. 6.8: TLR camera [5]

2. **Single-lens reflex (SLR) cameras**: The mirror and prism system is the backbone of the single-lens reflex camera (SLR). The photographer can see what will be captured

**Fig. 6.9 SLR** camera [6]

in real-time because of the camera's transparent lens. Pressing the shutter button on a mechanical SLR causes the mirror to flip out of the way, allowing light to reach the photosensitive sensor and the image to be captured.

3. **DSLR cameras**: Digital single-lens reflex (DSLR) cameras, as opposed to point-and-shoot models, offer superior image quality due to their larger image sensors. It allows for extensive manipulation of the shot by letting the user switch lenses and adjust the exposure settings (aperture, shutter speed, ISO, etc.).

**Fig. 6.10:** DSLR camera [7]

4. **Mirrorless Cameras**: Mirrorless cameras are an improvement on DSLRs because they use a digital display system instead of an optical one. The image sensor acts as a makeshift reflex mirror by relaying its signal to the viewfinder in the absence of a reflex mirror. In this way, the photographer can view the image in real-time on a tiny LCD screen located on the camera's back. This display serves as both the viewfinder and the control panel, eliminating the need for the latter.

### 6.5.3 Additional Accessories

Photography can be made efficient in terms of time and labour through the use of various cameras accessories which include:

1. **Tripod:** Tripods are extendable, three-legged stands typically used to mount cameras, with the foremost purpose of maintaining a steady and secure camera in a particular spot. Long exposures and taking multiple pictures from the same vantage point both necessitate it.
2. **Filters**: Filters, which are small pieces of glass fastened to lenses, help to filter the light that enters the lens. Therefore, it modifies the photo's hue, saturation, brightness, and contrast, and can even add some effects. Moreover, they protect the lens from debris, UV rays, and scratches.
3. **Flash**: The flash is the most popular piece of artificial lighting used in still photography, and it works by emitting a burst of light for a fraction of a second. Flashes are commonly built into modern cameras and when the shutter button is pressed during flash photography, the flash automatically activates to illuminate the subject. However, in order for external flashes to work in conjunction with the camera, a trigger cable or wireless trigger must be employed.

## 6.6 CAMCORDER

### 6.6.1 Introduction

With the help of a camcorder, you can record video and audio onto a mini cassette or DVD, or memory card. This is accomplished by combining the functions of a television camera, a video recorder, and a synchronization pulse generator. After that, the signals are reproduced on a standard DVD or another storage device. Microelectronics, IC design, and inventiveness in design have all combined to produce the digital camcorders of today. However, although the picture quality may not match with full-size professional equipment still is a good, cost-effective solution for personal usage. Furthermore, most of the time the user's lack of expertise rather than the camcorder itself is the cause for the substandard quality of the resulting images. The light-sensitive chip used in modern camcorders was first developed for satellite television. The digital system permits capturing stills along with some sound, and the compression greatly reduces the amount of space needed for storage.

Two innovations in the late '80s made it possible for camcorders to become much smaller. For instance, the VHS-C (Video Home System - Compact) videotape format was introduced. Which was compact and lightweight because the tapes could be played on any VHS player with the help of an adapter. Although the tapes could only hold up to 40 minutes of recordings, the devices' reduced size and weight made them useful in a wider range of situations. Moreover, the introduction of 8 mm tapes was the second major development for the video recording industry. Even though these tapes were much smaller than VHS tapes, they were still capable of holding two hours of high-quality video. However, a significant drawback was that they required a direct connection to a VHS or Betamax player in order to play or re-record footage.

**Fig. 6.11:** Camcorder [8]

### 6.6.2 Types of Camcorders

1. **MiniDV**: A digital tape format called MiniDV, which was the smallest tape ever developed up until that point, was widely adopted by camcorders by the end of the 1990s. Multiple copies could be made with little to no degradation in quality, and many formats can store more than three hours of video on a single tape. MiniDV is widely used in digital camcorders because it strikes a good balance between portability and cost.

2. **Digital8 camcorder**: Digital8 camcorders provide a novel digital alternative to standard video recorders. These digital tapes are compatible with both 8 mm and Hi8 systems, so they are an excellent option for those who wish to update from a lower-quality format. However, they cannot compare to the perks of MiniDV. Some new types of digital camcorders have lately become more popular due to falling prices of recordable media and portable storage,
3. **DVD camcorders**: DVD camcorders are DVRs that can also record to DVD-RAM or DVD-R discs for long-term storage. Due to its impending portability and low cost, the DVD format has seen an exponential rise in sales. Micro drives and flash memory-based mobile devices typically use flash memory or a specialised memory format to store video. Due to the high cost of storage for media larger than 2GB, flash-based camcorders are now a niche product.

## UNIT SUMMARY

Among the many digital electronic systems most widely used are fax machines, photocopiers, air conditioners, refrigerators, digital cameras, and camcorders. In order to send and receive faxes, a regular telephone line is used. However, a document must be digitised before it is modulated to sound for transmission. Moreover, to verify that the fax was successfully sent, some devices generate a confirmation page after sending.

The fundamental working principle of photocopiers is based on the law of attraction and the fact that some materials increase their electrical conductivity when exposed to electromagnetic radiation. Essential parts of any photocopier incorporate the photoreceptor drum, lamp and lenses, corona wires, fuser and toner which facilitate the photocopying procedure.

In order to improve the comfort of their users, air conditioners can adjust the relative humidity and temperature of a room by processing the refrigerant in the cooling cycle. There are many variations of the central air conditioner, but some of the most common is the mini-split, central ducted systems, central plant cooling, portable units, and packaged air conditioners.

Furthermore, the primary purpose of a refrigerator is to prolong the life of perishable foods by decreasing the growth of harmful bacteria. that's why it's been deemed essential for every home. The cooling process in a fridge is carried out by several different components, such as the refrigerant fluid, compressor, condenser, thermostatic expansion valves, evaporator, and receiver.

The term "camera" refers to any optical device that can record moving or still images and send or store them. Moreover, modern cameras, which evolved from the traditional camera, share many characteristics with the human eye. There is a wide variety of digital cameras, including point-and-shoot models, rangefinder models, single-lens reflex (SLR) models, digital single-lens reflex (DSLR) models, and mirrorless cameras. A camcorder combines the functions of a television camera, a video recorder, and a synchronisation pulse generator to record video and audio onto a mini cassette, DVD, or memory card and then reproduce the signals on a standard DVD or another storage device

## EXERCISES

### Multiple Choice Questions

1. Which of the following is not true for facsimile machines?
   (a) The majority of them employ a thermal printer
   (b) The width of the thermal printer is 8.5 inches
   (c) It includes a paper cutter
   (d) The height of the thermal printer is 8.5 inches
2. In a photocopier, __________ becomes charged when exposed to light.
   (a) Corona wires
   (b) Photoreceptor belt
   (c) Toner
   (d) Fuser
3. The refrigerant has high pressure and low temperature when it passes from ____
   (a) Evaporator to compressor
   (b) Compressor to condenser
   (c) Condenser to expansion valve
   (d) Expansion valve to evaporator
4. An ADC is used in a digital camera.
   (a) True
   (b) False
5. A camcorder is a combination of_______
   (a) Sync pulse generator, TV camera & video recorder
   (b) TV camera, video recorder & amplifier
   (c) Amplifier, sync pulse generator & video recorder
   (d) TV camera, sync pulse generator & ADC

| ANSWERS | | | | | | | | | |
|---|---|---|---|---|---|---|---|---|---|
| 1. | (d) | 2. | (b) | 3. | (c) | 4. | (c) | 5. | (a) |

### Short and Long Answer Type Questions

1. Briefly describe how to send a fax.
2. Explain the working of a photocopier.
3. State the different types of air-conditioning units.
4. Describe any 2 types of digital cameras.
5. Write a short note on a camcorder.
6. Explain in detail the workings of a fax machine. Draw the block diagram.
7. List and explain various components of a photocopier.
8. With the aid of a suitable diagram, explain how an air conditioner works.
9. State and describe the components of a refrigerator system.
10. Define and explain the laws that govern how a refrigerator functions.
11. Explain details of digital camera systems and their types.

## PRACTICAL

1. Study the working of a fax machine
2. Study the working of a photocopier and diagnosis basic operational faults.
3. Study the working of standard types of air conditioners and diagnosis of basic operational faults.
4. Study different components of a refrigerator system.
5. Study the working of a digital camera system and its types.

## KNOW MORE

- The International Energy Agency (IEA) predicts that by 2050, there will be 5.6 billion air conditioners in use throughout the world. According to estimates, this constituted about 20% of 2018's global building electricity consumption.
- In 1902, American inventor Willis H. Carrier put in place the first commercial air conditioning system at the Sackett-Williams Lithographing & Publishing Company. The Sackett-Williams office was in New York City.
- In 1959, Xerox introduced commercial xerographic office photocopying, which quickly began to displace older duplicating technologies like the Verifax, Photostat, carbon paper, and mimeograph machines. Office photocopying using commercial xerography was Xerox's big break.
- Carlson, the inventor of the photocopier, was arthritic and found the procedure of producing multiple duplicates of significant documents to be a time-consuming endeavour, so he decided to study photoconductivity.
- The Beta movie BMC-100P from Sony was the very first consumer camcorder, and it was released in 1983. But since it takes a Betamax cassette and sits on the operator's shoulder, it's not one that can be held with one hand.
- Kodak engineer Steve Sasson came up with the idea for the first digital camera in 1975 and built it out of scrap materials and unused kit components that were lying around the factory.
- First introduced in Japan in May 1999, the Kyocera Visual Phone VP-210 was the first mass-produced camera phone. The front-facing camera had 110000 pixels and could store 20 JPEG digital images for later transmission via e-mail or phone.

## REFERENCES


1. Fax machine, accessed 10 September 2022, <https://indianrenters.com/fax-machine-on-rent/>
2. Photocopier, accessed 10 September 2022, <https://www.collinsdictionary.com/dictionary/english/photocopier>
3. Refrigerator, accessed 10 September 2022, <https://www.mylloyd.com/refrigerator/direct-cool-refrigerator/gldf213sbbt2pb>
4. Point and Shoot camera, accessed 10 September 2022, <https://digital-photography-school.com/does-a-point-shoot-camera-still-make-sense-for-you/>

5. TLR camera, accessed 10 September 2022, <https://www.lomography.com/magazine/319470-8-tlr-cameras-for-the-creative-photographer>
6. SLR camera, accessed 10 September 2022, <https://flynngraphics.ca/the-collection/the-cameras/canonflex/>
7. DSLR camera, accessed 10 September 2022, <https://www.smartprix.com/cameras/dslr-type>
8. Mirrorless Camera, accessed 10 September 2022, <https://www.bhphotovideo.com/c/product/1659385-REG/canon_xf605_4k_pro_camcorder.html>
9. Sinclair, I. (2011). Electronics simplified. Newnes.
10. Frenzel, L. (2017). Electronics Explained: Fundamentals for Engineers, Technicians, and Makers. Newnes.
11. Bishop, O. (2001). Understand electronics. Elsevier.
12. Gates, E. (2011). Introduction to electronics. Cengage Learning.
13. Turner, L. W. (Ed.). (2013). Electronics engineer's reference book. Butterworth-Heinemann.
14. Gibilisco, S. (2001). The illustrated dictionary of electronics. McGraw-Hill Companies.
15. Eggleston, D. L. (2011). Basic electronics for scientists and engineers. Cambridge University Press.
16. Lancaster, K. (2018). DSLR Cinema: A beginner's guide to filmmaking on a budget. Routledge.

## REFERENCES FOR FURTHER LEARNING


1. Gupta, R. G. (2010). Audio & Video Systems, 2e. Tata McGraw-Hill Education.
2. Harris, K. D. (2007). U.S. Patent Application No. 29/224,646.
3. Gulati, R. R. (2005). Monochrome and colour television. New Age International.
4. Frenzel, L. (2017). Electronics Explained: Fundamentals for Engineers, Technicians, and Makers. Newnes.

## CO AND PO ATTAINMENT TABLE

Course outcomes (COs) for this course can be mapped with the programme outcomes (POs) after the completion of the course and a correlation can be made for the attainment of POs to analyze the gap. After proper analysis of the gap in the attainment of POs necessary measures can be taken to overcome the gaps.

### Table for CO and PO attainment

| Course Outcomes | Attainment of Programme Outcomes (1- Weak Correlation; 2- Medium correlation; 3- Strong Correlation) | | | | | | | | | | | |
|---|---|---|---|---|---|---|---|---|---|---|---|---|
| | PO-1 | PO-2 | PO-3 | PO-4 | PO-5 | PO-6 | PO-7 | PO-8 | PO-9 | PO-10 | PO-11 | PO-12 |
| CO-1 | | | | | | | | | | | | |
| CO-2 | | | | | | | | | | | | |
| CO-3 | | | | | | | | | | | | |
| CO-4 | | | | | | | | | | | | |
| CO-5 | | | | | | | | | | | | |
| CO-6 | | | | | | | | | | | | |

The data filled in the above table can be used for gap analysis.

# INDEX